'More Crop per Drop': Revisiting a Research Paradigm

Dedicated

to

Dr. Felix P. Amerasinghe

Our colleague and friend is no longer with us,
but his keen intellect, dedication, and lifelong contribution to the science and policy of water, health and environment continue to inspire all of us at IWMI.

‘More Crop per Drop’:
Revisiting a Research Paradigm

RESULTS AND SYNTHESIS OF IWMI’S RESEARCH, 1996-2005

Edited by

Meredith A. Giordano

Frank R. Rijsberman

R. Maria Saleth

International Water Management Institute

IWA Publishing

Published by IWA Publishing, Alliance House, 12 Caxton Street, London SW1H 0QS, UK

Telephone: +44 (0) 20 7654 5500; Fax: +44 (0) 20 7654 5555; Email: publications@iwap.co.uk
Web: www.iwapublishing.com

First published 2006

British Library Cataloguing in Publication Data
A CIP catalogue record for this book is available from the British Library

Library of Congress Cataloging- in-Publication Data
A catalog record for this book is available from the Library of Congress

ISBN: 1843391120
ISBN13: 9781843391128

CONTENTS

LIST OF TABLES

LIST OF FIGURES

LIST OF BOXES

FOREWORD

As populations rise, incomes grow, diets change, and countries industrialize, urban water demand will continue to increase apace. At the same time, greater environmental awareness will place more and more emphasis on maintaining healthy ecosystems for people as well as nature. Already large-scale development of river and groundwater resources has become less acceptable than it was from 1960 to 1990, when the large majority of the world's 45,000 large dams were built. Moreover, water infrastructure built during that period is becoming obsolete - through silting up of reservoirs and crumbling of irrigation networks - and there appears to be a decreasing willingness to fund rehabilitation and replacement of infrastructure. In addition, groundwater levels are falling in key aquifers that provide livelihoods and food security to millions of farmers.

One result of these changes is that access to water for agriculture is coming under increasing pressure. In the past, the agriculture sector grew based in part on access to cheap and plentiful water in irrigated areas. As the human population tripled in the twentieth century, water use multiplied sixfold, mostly for agriculture. The increase in agricultural productivity, in recent decades, has been due not only to higher-yielding varieties and increased fertilizer use but also to major investments in water resources infrastructure and massive energy subsidies for pumping groundwater - two phenomena that are less likely to be repeated in the coming decades. Moreover, agricultural growth made possible by irrigation does not always result in reduced poverty. As resources become scarcer and the negative consequences of additional development grow, the poor and vulnerable are impacted first and suffer most. Thus the question is: *How will we find sufficient water to provide food security, health, and livelihoods to a growing world population - in harmony with other water users and the environment?* This is truly a global challenge, and the focus of IWMI's work.

To grow enough food and provide sustainable livelihoods to poor people with the available water will require a considerable overhaul of the way agriculture is practiced. Based on its research, *IWMI believes it can make a key contribution to the water-food-environment challenge by demonstrating that increasing water productivity at the basin scale can contribute to improved food, health and livelihoods for poor people while sustaining the natural resource base.*

The concept of increasing water productivity ('more crop per drop') was first introduced at IWMI in the mid-1990s at a time when IWMI, then IIMI, undertook an in-depth synthesis and analysis of its first decade of research. The research synthesis, as well as the adoption of the 'more crop per drop' research paradigm, were documented in the volume *Expanding the Frontiers of Irrigation Management Research* (Merrey 1997). Since then the concept of water productivity has formed the cornerstone of IWMI's research program-whose key results, lessons and outcomes are presented in this second synthesis volume. In addition to documenting IWMI's past decade of research, we also use this opportunity to reflect upon the specific influences, lessons, and limitations of the 'more crop per drop' research paradigm. While we continue to place water (and land) productivity at the center of IWMI's research, we introduce in this volume a more refined notion of the concept.

We are proud of the achievements IWMI has made over the past decade and their emerging outcomes and impacts. As we learn from our past research and commence a new phase in IWMI's research program, we look forward to continuing our mission to improve the productivity of water and land resources for food, livelihoods and nature.

Frank R. Rijsberman
August 2006
Colombo, Sri Lanka

ACKNOWLEDGEMENTS

Most chapters included in this volume were initially prepared in 2004 and finalized through 2005. As a result, their coverage of IWMI research is comprehensive up to 2004, though many key works completed during 2005 are also covered.

The research synthesized in this volume would not have been possible but for the continuing support of IWMI's donors and partners (see annexes A and B, respectively). We are grateful to each one of them for their immense support and contributions. In the preparation of this volume, we would like to thank all the contributing authors for their intellectual inputs, constant cooperation, and prompt response.

Thanks are also due to IWMI researchers and external reviewers for their critical comments and suggestions on individual chapters in the book, including Drs. Jacob Kijne, Hugh Turral, Christopher Scott, Wim van der Hoek, Vladimir Smakhtin, Clifford Mutero, Eline Boelee, Robert Simmons, Stephanie Buechler, Jeroen Ensink, Flemming Konradsen, David Molyneux and several anonymous reviewers. We are indeed grateful to Prof. John Braden (University of Illinois at Urbana-Champaign) and Dr. Jerome Delli Priscoli (Institute of Water Studies, US Army Corps of Engineers and Editor-in-Chief, *Water Policy*) for their overall evaluation of the final manuscript.

We appreciate the excellent research and technical support extended by Sanjiv de Silva, Pavithra Amunugama, Janitha Godamuduna, Shalini Kumaresan, Arosha Ranasinghe, and Natalia Abeynayake. We thank Kingsley Kurukulasuriya for his excellent work on both copyediting and index preparation. Finally, we also thank IWA Publishing, in particular, Alan Click and Emma Gulseven, for their support and cooperation in bringing out this volume in such quality and in record time.

ACRONYMS AND SYMBOLS

AChE	Acetyl cholinesterase
ACIAR	Australian Council for International Agricultural Research
APO	Associate Professional Officer
AWDI	Alternate Wet and Dry Irrigation
B/C	Benefit/Cost
BOD	Biological Oxygen Demand
C	Committed
CA	Comprehensive Assessment of Water Management in Agriculture
CDE	Centre for Development and Environment
CGIAR	Consultative Group on International Agricultural Research
CMAs	Catchment Management Agencies
COD	Chemical Oxygen Demand
CPWF	Challenge Program on Water and Food
CSIRO	Commonwealth Scientific and Industrial Research Organization
DBL	Danish Bilharziasis Laboratory
DWAF	Department of Water Affairs and Forestry
DWFE	Dialogue on Water, Food and Environment
EC	Electrical Conductivity
EFR	Environmental Flow Requirements
ET	Evapotranspiration
EVI	Enhanced Vegetation Index
FAO	Food and Agriculture Organization of the United Nations
FDC	Flow Duration Curve
GDRS	General Directorate of Rural Services (Turkey)
GIS	Geographical Information System
GWI	Groundwater Irrigation
GWP	Global Water Partnership
H&E	Health and Environment
HRH	His Royal Highness
IAASTD	International Assessment on Agricultural Science and Technology Development
IAC	Inter Academy Council
IBIS	Indus Basin Irrigation System
IBSRAM	International Board for Soil Research and Management
ICID	International Commission for Irrigation and Drainage
ICLARM	International Center for Living Aquatic Resource Management (Now known as the WorldFish Center)
IDE	International Development Enterprises
IFAD	International Fund for Agricultural Development

IFAP	International Federation of Agricultural Producers
IFPRI	International Food Policy Research Institute
IGB	Indo-Gangetic Basin
IIMI	International Irrigation Management Institute
IMT	Irrigation Management Transfer
INRM	Integrated Natural Resources Management
IPM	Integrated Pest Management
IPTRID	International Program for Technology Research in Irrigation and Drainage
IRRI	International Rice Research Institute
IUCN	World Conservation Union
IWMA	Integrated Water Management for Agriculture
IWMI	International Water Management Institute
KOISP	Kirindi Oya Irrigation and Settlement Project
LAI	Leaf Area Index
m ha	million hectares
m mt	million metric tons
MDGs	Millenium Development Goals
MIA	Murrumbidgee Irrigation Area
MSEC	Management of Soil Erosion Consortium
NARES	National Agricultural Research and Extension Services
NB	Non-beneficial (e.g., water going to saline groundwater)
NDP	National Drainage Programmes
NGI	North Gujarat Sustainable Groundwater Management Institute
NPB	Non-Process Beneficial (forest evapotranspiration)
NWFP	North-West Frontier Province
O&M	Operation and Maintenance
P	Process fraction (crop evapotranspiration)
PEEM	Panel of Experts on Environmental Management
RO	Reverse Osmosis
SACEP	South Asia Cooperative Environment Program
SAR	Sodium Absorption Ratio
SDC	Swiss Agency for Development and Cooperation
SEI	Stockholm Environment Institute
SGM	Sustainable Groundwater Management
SGVP	Standardized Gross Value of Production
SIMA	System-Wide Initiative on Malaria and Agriculture
SIMI	Smallholder Irrigation Market Initiative
SOM	Soil Organic Matter
SWAP	Soil-Water-Atmosphere-Plant
SWI	Surface Water Irrigation
SWIM	System-Wide Initiative on Water Management

SWL	Static Water Level
SWNM	Soil, Water and Nutrient Management
TDS	Total Dissolved Solids
UNEP	United Nations Environment Programme
UNWWDR	United Nations World Water Development Report
WASA	Water and Sanitation Authority
WHE	Water, Health and Environment
WHO	World Health Organization
WOCAT	World Overview of Conservation Approaches and Technologies (Berne)
WRI	Water Resources Institute
WRIP	Water Resources Institutions and Policies
WWC	World Water Council
WWF	Worldwide Fund for Nature
$	US dollar/s

1

IWMI Research: Context and Setting

Meredith A. Giordano

1.1 INTRODUCTION

With increasing scarcity and competition for water at various scales, there is a growing public and policy demand for value-added information to better understand and address the issue of water scarcity and its food, livelihood and environmental implications. IWMI - in its quest to become a global knowledge center on water, food and the environment - generates and disseminates a wide variety of knowledge products, ranging from research reports to policy briefs for the international research, policy and donor communities. Besides these project-based outputs, IWMI also prepares synthesis volumes that review and summarize broader research programs and issues. Such synthesis work not only enhances the accessibility of IWMI's research results to a wider audience but also acts as a ready reference for research and policy purposes. Further, a comprehensive examination of IWMI's past works also provides an opportunity to reflect on the evolution and direction of the Institute's research focus over time and serves, therefore, as a basis for periodic reorientation of its research agenda and thematic priorities.

 More Crop per Drop. Edited by M.A. Giordano, F.R. Rijsberman, R. Maria Saleth.
ISBN: 9781843391128. Published by IWA Publishing, London, UK.

The present volume is one among such synthesis works, covering IWMI's research during 1996-04. It is a sequel to a similar volume by Merrey (1997) that covered IWMI's research during 1984-95 and documented well the major shift in the Institute's research focus from irrigation management at system level to water management at the basin scale. The present volume describes the evolution of IWMI's research agenda since that time and the further expansion of its mandate to encompass not only water but also land management issues and their larger implications for food, livelihoods and environment. This volume aims to provide an accessible, yet an in-depth and informative synthesis of the Institute's research on the key issues within what we refer to as the 'water-food-environment nexus'. Furthermore, since the volume covers a period that coincides with the development and application of a new research paradigm on water productivity (popularly known as 'more crop per drop') the thematic areas covered in the different chapters treat the topic as an analytical method as well as a means for dealing with water and related problems from a basin perspective.

The present chapter introduces the volume by setting the stage with a brief overview of the global water challenge and provides the rationale for a more integrated approach to research on water, food and environmental issues. Within this context, the chapter introduces IWMI's mission and the manner in which it has organized its research over the past decade to address this complex subject in an effective and meaningful way. Finally, this chapter provides an outline of the volume as a whole in order to inform the readers of the key topics and issues covered.

1.2 CHARACTERISTICS OF THE WATER CHALLENGE

Water experts have been engaged in a 'water crisis' discourse for several decades now with policymakers and the general public more recently taking greater interest in the topic. There is a growing consensus over the emergence of a water crisis both at the regional and global levels. However, opinions diverge on the nature of this crisis. The literature offers several perspectives.[1] For some, the crisis relates to *physical water scarcity*. Numerous water scarcity indices have been developed over the past 15 years to define water scarcity (e.g., Falkenmark 1986; Ohlsson 1999). These indices have been used to identify countries or regions at the greatest risk of water stress (Raskin *et al.* 1997; Montaigne 2002) and, extrapolating from that, for conflict (e.g., Klare 2001). Concerns over growing water scarcity have also prompted some to question whether there will be sufficient water for food production requirements as water

[1] Rijsberman (2006) reviews the relative validity of each of these perspectives both to place them in proper context as well as to develop a common understanding on the subject.

demands from the industrial and urban sectors expand. For others, the crisis relates essentially to the lack of access to the resource. The World Health Organization, for example, estimates that 1.1 billion people are currently without access to improved water supply sources and, more than 2.4 billion people lack adequate sanitation (WHO 2003). Similarly, reliable and affordable access to water for food production is out of reach for many of the world's 900 million rural poor (Rijsberman 2004). For these cases, the problem is not so much related to the nonavailability of the resource per se, but rather to the lack of political will and financial resources to make water available for the consumptive and productive needs of the poor and unserved communities. Here, water scarcity involves equity considerations with economic and political economy dimensions.

An *environmental water crisis* is yet another manifestation of unsustainable water extraction and use, whose socioeconomic implications are as large as its ecological consequences. It is caused by declining quantity and quality of water available for ecological systems covering the spectrum from upstream forests to downstream wetlands and with critical significance for the livelihoods of the poor. Water-starved ecosystems are clearly an outcome of intensive water withdrawals for irrigation and urban purposes, increasing pollution, and improper land use practices. It is estimated, for example, that half of the world's wetlands have been lost to agricultural development (UNWWDR 2003), and the construction of dams has led to the destruction of 25 million kilometers of riverine systems (Cosgrove and Rijsberman 2000). Finally, taking all these issues into account, many argue that the true source of the crisis is not the physical scarcity per se but the *lack of proper management* of the water and land resources both at local and global levels (e.g., Cosgrove and Rijsberman 2000; World Bank 2003; Rogers and Hall 2003). It is based on such a diagnosis that various international fora - from the 1992 Dublin Conference to the 2003 Third World Water Forum - manifestly identify policy and institutional reforms for effective water governance as the highest priorities for action.

1.3 WATER-FOOD-ENVIRONMENT NEXUS

The true nature of the water crisis, however defined, and its causes and consequences are clearly complex. The challenges involve sufficient supplies of water for food production, improved access to productive land and water resources for the world's poor, and minimizing the trade-offs between agriculture and the environment. In short, we refer to this complex set of issues as the *water-food-environment nexus*. Identifying appropriate solutions to the challenges requires both a detailed understanding of this nexus as well as an assessment of possible response mechanisms, both technical and institutional, and their related impacts across multiple scales and sectors.

For many regions of the world, increasing water productivity (or 'more crop per drop') of irrigated and rain-fed systems, rather than allocating more water, holds the greatest potential to improve food security and reduce poverty with the least environmental cost. For example, research suggests that improving water productivity by 40% on rain-fed and irrigated lands can reduce the need for additional withdrawals for irrigation to zero over the next 25 years. While the irrigation systems in Europe, the US, China and Brazil are already operating at high water productivity levels, there is great scope for achieving productivity and related livelihood gains in other regions, particularly in Africa and Asia (Molden and de Fraiture 2004; Cai and Rosegrant 2003; Rockström *et al.* 2003). However, simultaneously tackling issues of food production, poverty alleviation and environmental sustainability requires a broader definition of water productivity than that implied by the slogan '*more crop per drop*' with its focus on crop yields alone. Water productivity needs to be understood in the widest possible sense so as to account for the full range of benefits from water use, including crop yields, land and soil fertility, fishery outputs and ecosystem services as well as the associated social benefits such as improved health and nutrition. Furthermore, its implications must be understood not just at the farm and field levels but also at basin scales and across sectors. Simply stated, the challenge is to catalyze effective and efficient improvements in water productivity at the basin scale in a way that simultaneously achieves food security, poverty alleviation and environmental sustainability goals.

1.4 IWMI'S RESEARCH RESPONSE

For nearly 20 years, the International Water Management Institute (IWMI), formerly the International Irrigation Management Institute (IIMI), has worked with its partners in developing countries to improve the management of water and land resources for agriculture through better technologies, policies, institutions and management. A member of the Consultative Group on International Agricultural Research (CGIAR) since 1992, IWMI endeavors to bring together researchers and practitioners to identify practical solutions to water-related problems in agriculture. Initially focused on irrigation, the Institute has since broadened its mandate to examine the management of both water and land, two closely connected resources, with the vision of increasing food security and improving the health and livelihoods of the world's poor while protecting the surrounding environment.

Since the mid-1990s, IWMI's research has focused primarily on opportunities to improve the productivity ('more crop per drop') of water for agriculture at the basin scale. As such, the key research question for the Institute has been: "*How can we grow more food and sustain rural livelihoods with less water in a manner that is socially acceptable and environmentally sustainable?"* To respond to this question, IWMI's research agenda over the past decade was organized around five thematic areas, namely:

(1) Integrated Water Resources Management.
(2) Smallholder Land and Water Management.
(3) Sustainable Groundwater Management.
(4) Water Resources Institutions and Policies.
(5) Water, Health and Environment.

In addition to these five themes, since 2000 IWMI has also led a 6-year, multi-institute initiative called the Comprehensive Assessment of Water Management in Agriculture (CA, for short).[2] The CA takes stock of how water for agriculture has been managed over the last 50 years and the impact of the past policies and practices on food and environmental security. This volume includes a summary and synthesis of the past research in the five thematic areas noted above as well as the emerging results from the CA.

1.5 CONTEXT AND ORGANIZATION OF THIS VOLUME

The volume is planned essentially as an analytical summary and a critical synthesis of research at IWMI over the past decade under its evolving research paradigm of 'more crop per drop'. As this period coincided with a broadened mandate for IWMI, i.e., from 'irrigation management' to 'land and water management', the research work synthesized in this volume covers the full range of issues falling on the larger canvas of the water-food-environment nexus.

The volume is organized around nine chapters. In the chapter that immediately follows this introduction, Frank Rijsberman provides a critical review of the 'more crop per drop' research paradigm and its implications for IWMI's current and future research. He presents the evolution of the paradigm, including its influence and limitations, and how the lessons learned from a decade of research on the topic are now influencing IWMI's newly revised research agenda. Within this context, Rijsberman provides the rationale for rethinking the 'more crop per drop' research paradigm and the associated refinement of IWMI's research agenda. The next five chapters present the review and synthesis of research under the aforementioned five core thematic areas and the CA.

In chapter 3, Hammond Murray-Rust and Hugh Turral review and summarize the research work under the theme Integrated Water Resources Management. Having discussed the logic and rationale for selecting river basins

---

[2] The CA was formed in 2000 out of the former CGIAR System-Wide Initiative on Water Management (SWIM) and is thus also referred to as SWIM-2.

rather than individual irrigation systems as a unit of analysis, the authors describe the approaches and methodologies that IWMI has developed and applied for basin-level evaluation of water scarcity, water productivity, water accounting and related performance indicators. The chapter also presents a series of case studies in river basins throughout Asia and Africa. Chapter 4, contributed by Frits Penning de Vries and Deborah Bossio, describes the integration of land management into IWMI's research agenda and summarizes the related research on the land-water interface with particular focus on the livelihoods of smallholders and the economic and environmental sustainability of degraded lands and catchments. Using case studies from Asia and Africa, the chapter presents results on the livelihood and ecological effects, both on- and off-site, of improving water access and productivity among smallholders, promoting catchment conservation and reversing land degradation.

Chapter 5, contributed by Tushaar Shah, reviews the research under the theme of Sustainable Groundwater Management. Using a unique framework of groundwater socio-ecology, the author presents results on a wide variety of issues ranging from the welfare and productivity effects of groundwater irrigation to the conjunctive use and recharge options for resource management. The chapter also presents multi-country experiences with water and energy pricing, direct regulations and water-saving technologies. Field results of IWMI research on groundwater issues focusing on India, China, Mexico and South Africa are also provided. Chapter 6 by Madar Samad summarizes the research under the theme of Water Resources Institutions and Policies. It provides a concise review of research in five broad areas: institutional reforms in the irrigation sector; institutional analysis for river-basin management; water-poverty linkages; gender issues in irrigation; and economic issues, particularly water pricing and investment strategies. It presents not only their conceptual, analytical and methodological aspects but also the empirical insights from their applications in the context of several countries of Asia, Africa and Latin America.

Chapter 7, contributed by Felix Amerasinghe, provides a review and summary of research under the theme Water, Health and Environment. Against a brief note on the evolution of research in the water-health-environment interface at IWMI, this chapter synthesizes the research in five key areas: the health impacts of the irrigation-malaria nexus, the health and livelihood effects of wastewater-based agriculture, the extent and impacts of multiple water use, the economic and ecological effects of using basin catchments and wetlands and the health hazards of farm pesticides. This chapter presents field-based results - both from IWMI projects and other works -from a wide variety of countries in Asia and Africa. In chapter 8, David Molden provides the initial results from the ambitious multi-organizational research initiative of the CA. This chapter provides a synthetic overview of results in terms of some key research questions such as the magnitude of water needed to meet food demand; the benefits, costs and impacts of irrigation; the options for improving water productivity in both irrigated and rain-fed agriculture; and the institutional and policy options for balancing food and environmental water needs. Chapter 9 concludes the volume by bringing forth the major insights of research under different themes, highlighting some of the key outcomes and impacts of IWMI research on local water policies and programs as well as on global water research and policy debates.

As can be seen from the outline above, the volume describes new tools, approaches and methodologies and also illustrates their practical application both from a global perspective as well as in the local and regional contexts of Asia, Africa and, to some extent, Latin America. Since this volume brings together all major research of IWMI over the period 1996-04, including an almost exhaustive list of citations, in one single set of pages, it is expected to be valuable as research and reference material, as a policy tool and as a general source of information. Ultimately, it is hoped that by disseminating the major findings and key policy insights among the research, donor and policy communities, this volume will not only provide a capstone for IWMI's last 10 years of work but will also foster a new way of looking at the water issues within the broader development context of improving food production, rural livelihoods and human and environmental health.

2

'More Crop per Drop': Realigning a Research Paradigm

Frank R. Rijsberman

2.1 INTRODUCTION

The idea of 'more crop per drop' has been very influential both as a conceptual framework and as a guiding principle for organizing research within IWMI for over a decade now. It has emerged as a consequence of the broader mandate of the organization set in 1996 when the research focus broadened from irrigation management to agricultural water management. It also represents a fundamental change in the Institute's research paradigm, from one focused on 'irrigation efficiency' and 'system performance' to the one centered on 'water productivity' and 'basin management'. From a larger perspective, the broadened mandate and research focus of IWMI is, in fact, a logical response to the changing challenges of the global water sector where physical and economic scarcities of the resource increasingly affect the basins in developing regions with a high population pressure and a heavy dependence on water for food production and livelihoods.

The 'more crop per drop' paradigm provides a basis for solving many water-related problems by applying a 'soft path' of increasing overall water productivity, especially in the agriculture sector, in order to free up water for

 More Crop per Drop. Edited by M.A. Giordano, F.R. Rijsberman, R. Maria Saleth.
ISBN: 9781843391128. Published by IWA Publishing, London, UK.

other productive uses, including environmental water needs.[1] The idea is simple enough to catch the attention of the public, yet sufficiently complex to apply in research and as a basis for public policy. If interpreted in its literal sense, the focus is only on producing more crop output from a given or reduced water use. In reality, however, improving water productivity involves the maximization of social, economic and ecological services to society as a whole. To apply and assess this broader notion of water productivity are clearly much more difficult in view of the inherent conceptual, methodological and informational challenges. It is precisely this challenge that IWMI is trying to address through its basin-oriented research program.

This chapter aims to describe the evolution of the past and present research agenda of IWMI and the research and operational influences of the 'more crop per drop' paradigm. It also takes a critical look at the 'more crop per drop' idea and provides the rationale for adjusting the research paradigm in line with the broader challenges of the water-food-environment nexus. The chapter concludes with the delineation of the new conceptual framework and thematic structure for IWMI's research that are expected to assure a more effective application of the realigned research paradigm centered on 'more crop per drop'.

2.2 EVOLUTION OF IWMI'S RESEARCH PARADIGM

In 1996, David Seckler, then the recently appointed Director General of IWMI with a mandate to refocus the research agenda of the Institute, published the first IWMI Research Report (Seckler 1996). This brief note of about 10 pages contained many of the basic ideas that have come to characterize what has been coined as the IWMI approach to water for agriculture. It was in essence a research agenda around the following three ideas:

(a) *Basin focus*: As the degree to which the available renewable water resources in a river basin approach the maximum, and competition among users increases, the appropriate focus for water management is the basin level, not the field, farm or even irrigation system level; this basin concept is closely linked to the idea of open, closing and closed basins-where a basin is defined as closed when there is no usable water leaving the basin.

(b) *Recycling*: Many of the water savings achieved at field level may only capture water that would otherwise have been reused downstream; these are not real water savings, where additional supplies become

[1] Many researchers (e.g., Gleick 2003) now call for the "soft path for water" using the term of "soft path" coined originally by Amory Lovins for the energy sector. A similar approach has likewise been suggested in various IWMI publications over the last 7–8 years (e.g., Kijne *et al.* 2003; Rijsberman 2004).

usable for an additional use ('wet' water savings), but they are simply a reallocation of water from downstream to upstream users ('dry' water savings); with this idea comes a focus on the fate of water through recycling and reuse.

(c) *Crop water productivity*: Rather than focusing on the potentially misleading idea of increasing irrigation efficiency, the focus should be on increasing water productivity.[2] This, in essence, captures the output produced per unit of water consumed and hence, it is implicit in the idea of 'more crop per drop'.

These ideas, which are described in more detail in chapter 3, formed the core of the IWMI research agenda during 1996-00 and culminated in a number of significant publications, including the key work on water productivity (Kijne *et al.* 2003).

IWMI's focus on the basin level is closely linked to the question of water scarcity. In all but the driest areas of the world, the water-related development effort in agriculture has been focused on investments in infrastructure to make water available to meet the rising demands, i.e., a 'supply focus'. Among water professionals, the discussion since about the 1980s has been increasingly focused on approaches to managing demands to live within the means of finite supplies. IWMI's research has conceptualized that there are three different stages in water resources development, i.e., 'development' 'utilization' and 'reallocation', which are all closely related to the share of usable water supplies of the basin that has already been developed (Keller *et al.* 1998; Seckler *et al.* 1998a; Molle 2003; Molden *et al.* 2005). These stages are shown in Figure 2.1. IWMI has, therefore, argued that the nature of water resources management changes with the degree of water resources development in a basin. The bottom line is that when basins are closed or closing with no or low scope for developing additional water supplies, additional investment in water infrastructure can only shift water from one user/use to another; it does not produce additional water (when aggregated at basin level).

The primary conclusion of IWMI's work during the 1996-00 period, as presented in a widely cited IWMI Research Report (Seckler *et al.* 1998a) and later summarized by Seckler *et al.* (2003), aimed to increase awareness of impending water scarcity:

---

[2] While IWMI authors use "water productivity" consistently, other authors also use "water use efficiency" (e.g., Wallace 2000) to denote the same concept of output over water consumed. This is in contrast to the various definitions of "irrigation efficiency" that all indicate the share of water "used" as a percentage of the total applied—suggesting that the remainder is "lost" (while it is often reused), See, for instance, Seckler *et al.* 2003 for a review.

"[O]ne-third of the population lives in regions that have absolute water scarcity, in the sense that they do not have sufficient water resources to meet their agricultural, domestic, industrial and environmental needs in the year 2025...an additional 500 million people live in regions of severe economic scarcity; they have a sufficient amount of potential water resources to meet their 2025 needs, but they will have to more than double their present utilization of these resources through large, expensive and possibly environmentally destructive development projects..."

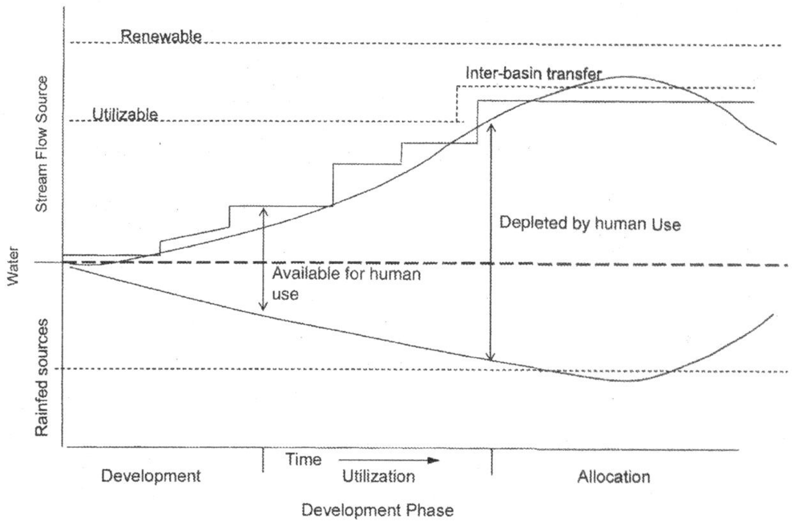

Figure 2.1 Phases of river basin development. Source: Molden *et al.* 2005, p.22.

Among water professionals, there has been much talk of a 'global water crisis' for several decades. Many would take the second World Water Forum, in the year 2000, as the moment where over 120 ministers, over 5,000 stakeholder representatives and water professionals, and over 600 journalists definitively put water on the map as a 'major global issue' (HRH the Prince of Orange and Rijsberman '2000).

IWMI was a key contributor to this process; the 'basic IWMI scenario' was published as the IWMI contribution to the World Water Vision (Cosgrove and Rijsberman 2000). The major Findings and Recommendations contained in this scenario were (IWMI 2000):

(a) The world's primary water supply will need to increase by 22% to meet the needs of all sectors in 2025.

(b) 17% more irrigation water will be needed for the world to feed itself in 2025.

(c) Nearly one-third of the populations of developing countries in 2025, some 2.7 billion people, will live in regions facing severe water scarcity.

(d) The global community must invest in research to improve crop water productivity (more crop per drop).

(e) New water infrastructure will have to be developed to meet future food requirements.

(f) Groundwater reserves will be increasingly depleted in large areas of the world.

(g) Salinization of soils, compounded in many cases by increasingly saline or poisoned groundwater, will seriously affect land that has been highly productive in recent decades.

(h) The people most affected by growing water scarcity will continue to be the poor, especially the rural poor; and among the poor people, women and children will suffer the most.

(i) Better use of water in several large internationally shared river basins can contribute significantly to achieving food security and reducing poverty in developing countries.

Seckler believed the solution to the water scarcity issues was to improve crop water productivity in irrigated agriculture as much as possible, but further development of water supplies for irrigation to meet future food demands was inevitable and would require the widely-cited *"17% of additional water for irrigation by 2025"*. He did not, however, believe in the potential to significantly improve water productivity in rain-fed agriculture. Indeed, the assumed low growth in water productivity in rain-fed agriculture in the 'basic IWMI scenario' is a key factor in the relatively high estimate of 17% growth in irrigation water demands.

2.3 INFLUENCE OF 'MORE CROP PER DROP' IDEAS

The ideas IWMI developed and promoted-often referred to as the 'more crop per drop' paradigm-have been very influential. A key conclusion drawn at the second World Water Forum in the key policy document discussed there, the World Water Vision, was that the nature of water scarcity, for the world as a whole, is not that the world is running out of water, but that we are managing it so badly that many people and the environment already suffer (Cosgrove and Rijsberman 2000). The Global Water Partnership concluded, *"on the one hand, the fundamental fear of food shortages encourages ever greater use of water resources for agriculture. On the other, there is a need to divert water from* *irrigated food production to other users and to protect the resource and the ecosystem. Many believe this conflict is one of the most critical problems to be tackled in the early 21st century"* (GWP 2000, p.58). In the same year, the UN Secretary General, in his report to the Millennium Conference, concluded, "*We need a Blue Revolution in agriculture that focuses on increasing productivity per unit of water - more crop per drop*" (Kofie A. Annan 2000).

IWMI[3] used the same arguments and, given the growing support for them in academic as well as in policy circles, they were also used as a basis to initiate three interrelated, major international initiatives with significant inter-organizational research and operational collaborations both within and outside of the Consultative Group on International Agricultural Research (CGIAR) system:

(a) Dialogue on Water, Food and Environment.
(b) CA.
(c) Challenge Program on Water and Food (CPWF).

The Dialogue on Water, Food and Environment (DWFE) is a strategic alliance involving ten key stakeholders in the water, agriculture and environment areas. It aims to help bridge the chasm between the agricultural and environmental communities over the way water should be developed, allocated and managed. These organizations ranged from UN agencies (FAO, UNEP and WHO) to associations of farmers (IFAP), irrigation engineers (ICID), environmental organizations (IUCN, WWF), water umbrella organizations (GWP, WWC) and water research (IWMI, representing the CGIAR). IWMI provided the Chair of the Dialogue Consortium and hosted the Secretariat. The Dialogue was organized around three main groups of activities:

(a) Promotion of cross-sectoral dialogues at national and basin levels, organized by the national committees/members/associations/offices of, for example, ICID, IUCN, IFAP, GWP and WWF.
(b) Creation of a 'knowledge-base' of credible and mutually agreeable information for both the agricultural and environmental communities based largely on linking and adding to the knowledge already available within the CGIAR and the United Nations systems as well as in international organizations such as the World Bank.
(c) Organization of local-action activities that aim to provide an information exchange and best-practice identification platform, linking thousands of local, NGO and bilateral projects and activities into a formal knowledge system.[4]

---

[3] Even though Seckler left IWMI in 2000, this author, as its new Director General, pursued largely the same agenda.

[4] See Bhatt and Vallee 2004 and Vallee 2005 for further details on the Dialogue on Water, Food and Environment.

The CA was developed in parallel with the Dialogue. It is a $15 million research effort that has brought together at least 300 researchers for an assessment similar in scale and spirit to the two international assessments of the global impacts of climate change and ozone depletion. The assessment is cosponsored by the CGIAR, FAO and the Ramsar Convention on wetlands. It is also positioned as the major water-input for the new International Assessment on Agricultural Science and Technology for Development (IAASTD). While the major findings of the CA will be formally disseminated in 2006, several key findings are already beginning to shape the nature of water and land management research. For example, the CA has made important contributions to understanding the environmental consequences of irrigated agriculture through the development of a global framework for assessing environmental flow requirements (Smakhtin *et al.* 2004) and analyzing the negative *and* positive externalities associated with an irrigated landscape (Galbraith *et al.* 2005; Bambaradeniya and Amarasinghe 2003). Additionally, the CA has contributed to a greater understanding of irrigation and poverty (e.g., Matsuno *et al.* 2002; Bhattarai *et al.* 2003; Boisvert *et al.* 2003; Hussain and Hanjra 2003) and is making significant strides in assessing key options to change the deeply rooted traditional approach implied in the 'more food = more water' equation, including opportunities to increase water productivity of irrigated and rain-fed systems (e.g., Barker *et al.* 2001; Dong *et al.* 2001; Inocencio *et al.* 2003; Namara *et al.* 2003).[5] A full summary of the key CA findings to date is provided in chapter 8.

In October 2001, at the CGIAR annual meeting, IWMI called for a major new CGIAR program to address the water crisis through agricultural research (Rijsberman and Molden 2001). It was argued that to solve "the 'world water crisis' in a major way", the challenge is to grow more food with less water - decreasing water use in agriculture to meet environmental goals and other human needs, yet growing enough food and improving livelihoods of the poor. IWMI estimated that over a 25-year period, a 60% increase of water productivity in irrigated lands, and a 30% increase of the same in rain-fed lands would be required as a major step in the right direction (see Table 2.1). This also marked the occasion where IWMI started arguing for a broader interpretation of water productivity than 'more crop per drop' alone. Water productivity, it said, needs to be understood in the widest possible sense - including crop yields, fisheries, ecosystem services and direct social benefits such as to public health. The challenge for the CGIAR was to catalyze effective and efficient improvements of water productivity in a way that is pro-poor, gender-equitable and environmentally sustainable.

---

[5] See chapter 8 for more details on the CA.

Table 2.1 Growth rates of water productivity and cereal yield needed to achieve food and environmental security.

Growth rates	Irrigated (%)	Rain-fed (%)
Recent annual growth in cereal yield	1.0	0.5
Business as Usual scenario		
Annual growth in cereal yield	1.0	0.5
Annual growth in water productivity	0.6	0.5
Total growth in water productivity (25 years)	20	15
Food and Environmental Security scenario		
Annual growth in cereal yield	1.3	1.0
Annual growth in water productivity	1.8	1.2
Total growth in water productivity (25 years)	60	30

Note: The 'Business-as-Usual' scenario forecasts an increase in water resources withdrawn for agriculture by 12-17% from 2000 to 2025. The 'Food-and-Environmental-Security' scenario would reduce the total withdrawal for agriculture by 10% for the period 2000-25.

Source: Rijsberman and Molden 2001.

One year later, in October 2002, the CGIAR approved a first phase of the CGIAR CPWF that involves an inception year and 5 years of research. Thanks to the growing recognition of its urgency and importance by international donors and development agencies, this program, at present, has a target budget of some $80 million for the first phase. Clearly, this is an attainable target, as by 2005, the CPWF is already implementing over 30 projects with a total budget of some $60 million in nine major river basins around the world: the Andes,[6] Indo-Gangetic, Kharkhe, Limpopo, Mekong, Nile, São Francisco, Volta and the Yellow. We estimate that these two research programs - the CA and CPWF - together are now engaging the participation of over a thousand scientists to become the flagship programs for global research on water, agriculture and development.

2.4 'MORE CROP PER DROP': A CRITICAL LOOK

IWMI research initiatives centered on the 'more crop per drop' paradigm have certainly been very influential at the international level. But, IWMI did recognize some of the limitations of this paradigm. As will be indicated below, these limitations have more to do with the restrictive interpretations and applications of the idea than with the basic concept on which it is based. There is also a tendency to confuse this paradigm as an end in itself, contrary to the fundamental need of treating it only as a means to address the larger problem of

[6] This is a grouping of smaller basins in the Andean region.

water reallocation. The specific limitations of the paradigm and the ways they are being amended through a refined research framework are discussed below.

2.4.1 Limitations of the paradigm

First, while the paradigm places a major emphasis on water reuse, it underplays the implications of such reuse for water quality and economic costs. The emphasis on the potential reuse of the fraction of water that is not consumed at different stages of the production process appears to suggest that such reuse can take place without a cost. But, this is not true as virtually all water withdrawals and applications lead inevitably to both private and social costs in the form of higher energy and application costs and water-quality degradation (salinization and pollution).

Second, the 'more crop per drop' concept does not accommodate the non-crop water outputs with considerable income, livelihood and health implications. These outputs include fisheries, environmental services and other multiple benefits gained from domestic water use to livestock watering. The implication is that while the focus on crop water productivity at farm or field level can often be justified, at larger scales, a broader definition of water productivity is needed that incorporates all values associated with water use. Only such a broader definition will serve the management of water across the many uses within a basin.

A third important limitation is the implicit emphasis on applied water for irrigation from renewable water resources, i.e., the part of the water cycle that runs off into rivers and recharges groundwater (also called blue water). Such a restrictive emphasis on blue water, however, tends to underplay the importance of the other 60% of the hydrological cycle that is stored as soil moisture (the so-called green water), which is the mainstay for rain-fed cultivation. With the growing importance of groundwater irrigation, small-scale irrigation, rainwater harvesting and supplemental irrigation, the sharply held boundaries between rain-fed and irrigated agriculture are now rapidly disappearing. This changed condition obviously requires a new, broadened and unified approach for evaluating water productivity across the whole spectrum of the hydrological cycle ranging from rain-fed to the irrigated systems.

Fourth, the objective of increasing water productivity, which is central in the 'more crop per drop' paradigm, is only a means but not the end in itself. The ultimate goal is the reallocation of water saved in the process to achieve the reduction of poverty and hunger while sustaining food, livelihood and environmental security. While increasing water (and land) productivity can well be a key factor for poverty alleviation and livelihood generation, this effect cannot be taken as automatic without first ensuring the poor to have access to productive land and water resources. Similarly, given the income, livelihood and health importance of many water-based ecosystems, there is also an urgent need to meet the growing water needs of the environment, particularly to support riverine and wetland ecosystems. Therefore, the focus on water productivity should not divert our attention from the fundamental issues of resource access and allocation as well as their distributional and environmental effects.

Finally, the paradigm is silent on the issue of sustainable use of natural resources, especially on the question of how to arrest resource overuse and degradation (e.g., groundwater depletion, water pollution, salinization, nutrient mining and soil erosion). Increasing water productivity is unlikely to halt overuse of water locally, as it is likely to increase the profitability of the farmer whose productivity has increased. In fact, it can encourage increased, rather than decreased, resource use. On a larger scale, assuming that the total demand for a given good or service stays constant, increased productivity in one location ought to displace water use with a lower productivity elsewhere. But, such a spatial reallocation, though good from a productivity perspective, may not necessarily increase the sustainability of water and land resource use in the basin or subbasin where the productivity has increased.

2.4.2 Addressing the limitations

During the period 2000-05, IWMI tried to address several of the major limitations associated with the 'more crop per drop' idea. This has involved major adjustments in, and additions to, IWMI's research agenda. The most important among these changes are listed below.

First, the balance between water for food and water for nature has become the core issue on the agenda, making the issues of access and allocation central to the Institute's research agenda. With this, IWMI has refocused its research around the so-called water-food-environment nexus (Rijsberman and Molden 2001; Rijsberman and Mohammed 2003; Rijsberman and de Silva 2004).

Second, the balance between productivity and sustainability is also restored with major emphasis on resource sustainability. As a result, the linkages between water and land, salinization and soil degradation, water and land quality, and nutrient cycling and reuse of wastewater in peri-urban agriculture have become a central focus of IWMI's work (Scott *et al.* 2004).

Third, IWMI has made a concerted effort to broaden the concept of water productivity to include major non-crop benefits and to consider its application in more contexts beyond irrigated agriculture. As a result, IWMI's research is increasingly covering non-crop benefits such as those from fisheries, livestock, wetlands, multiple use systems and biodiversity conservation (e.g., Nguyen-Khoa *et al.* 2005; Puskur and Thorpe 2005; Boelee and Laamrani 2004). Similarly, improving water productivity across the entire blue-green, rain-fed-irrigated and surface-groundwater spectra has become the norm in the work of the Institute (e.g., Noble *et al.* 2004; Kumar and Singh 2005; Qureshi *et al.* 2004). Notably, this has also led to a reassessment of the potential to improve water productivity in rain-fed agriculture, which was a serious gap in past applications of the 'more crop per drop' idea.

Finally, with explicit recognition of water productivity as only a means to achieve the larger goals of poverty alleviation and resource sustainability, attention is now more focused on the assessment of the impacts of water productivity on poverty alleviation, livelihood generation and environmental sustainability (e.g., Saleth *et al.* 2003; Smakhtin 2003; Scott *et al.* 2004; Puskur and Thorpe 2005).

Although these adjustments are essentially in the nature of both broadening and generalizing the concept and its application, they were fundamental in making the research paradigm more realistic and forward-looking. As will be shown in the next section, these adjustments were also very important in view of their far-reaching implications for the reorganization of the research agenda and thematic structure of IWMI.

2.5 THE WAY FORWARD

Initially, IWMI attempted to adjust and apply the 'more crop per drop' paradigm by organizing its research around five core themes as specified in its 2000-05 Strategic Plan. These themes are:

Theme 1: Integrated Water Resources Management.
Theme 2: Smallholder Land and Water Management.
Theme 3: Sustainable Groundwater Management.
Theme 4: Water Resources Institutions and Policies.
Theme 5: Water, Health and Environment.

The research under these themes and the CA are reviewed and synthesized in the following six chapters.

Based on the lessons learned over the past 10 years of research at IWMI, the results emerging from the CA, and the comments received through external reviews and stakeholder surveys as part of IWMI's recent 2004-08 Strategic Planning exercise,[7] IWMI has further refined its research framework and tightened its thematic structure. Thus in early 2005, IWMI reorganized its research framework and agenda to more directly address the issues of water and land productivity and possible interventions and related impacts. Briefly, the framework organizes IWMI's research around four activities:

---

[7] The 2004-2008 IWMI Strategic Plan is available on the IWMI website at: http://www.iwmi.cgiar.org/about/stratplan/Strategic%20Plan%202004-2008.pdf

(a) *Mapping water productivity:* This activity assesses water (and land) productivity at the basin level for key crops, combinations of crops, complementary livestock/fishery enterprise outputs, specific livelihood strategies, and environmental uses and values. To make the exercise more meaningful and practical, the assessments will also be performed at a spatially disaggregated level so as to analyze the key variables that explain the variations in water productivity (including soil/land degradation) across a basin. The key idea is not to suggest that water productivity is a solution, but rather as a valuable framework for understanding the productive uses of land and water resources.

(b) *Mapping water poverty:* This activity aims to assess the spatial patterns of poverty and access of poor people to productive land and water resources throughout the basin. The basic idea is not to presume that increasing water productivity will alleviate poverty, but rather to identify the target group that could benefit from improved access to productive land and water resources.

(c) *Analyzing high-potential interventions:* This activity will identify, assess and develop interventions (e.g., technologies, practices, and institutions and policies) that can increase water and land productivity, enhance the access of the poor to productive water and land resources, and improve the sustainability of resource use.

(d) *Assessing the impacts:* This activity aims to assess the potential impacts of interventions on water and land productivity, water poverty, livelihoods, health and the sustainability of the resource base under different adoption scenarios, knowledge-sharing models and developments in exogenous variables. Such assessments for different interventions will be carried out both at the basin and subbasin scales.

With their logical and operational linkages, these four activities define together a more realistic conceptual framework for research (see Figure 2.2). Added to this refinement of research framework, IWMI has also reorganized its research agenda to sharpen the focus more directly on the central issues of water and land productivity and their impacts on poverty and environment. The new research themes that will direct future research at IWMI are:

Theme 1: Basin Water Management (*understanding water productivity*).
Theme 2: Land, Water and Livelihoods (*improving livelihoods for the rural poor*).
Theme 3: Agriculture, Water and Cities (*making an asset out of wastewater*).

Theme 4: Water Management and Environment (*balancing water for food and nature*).

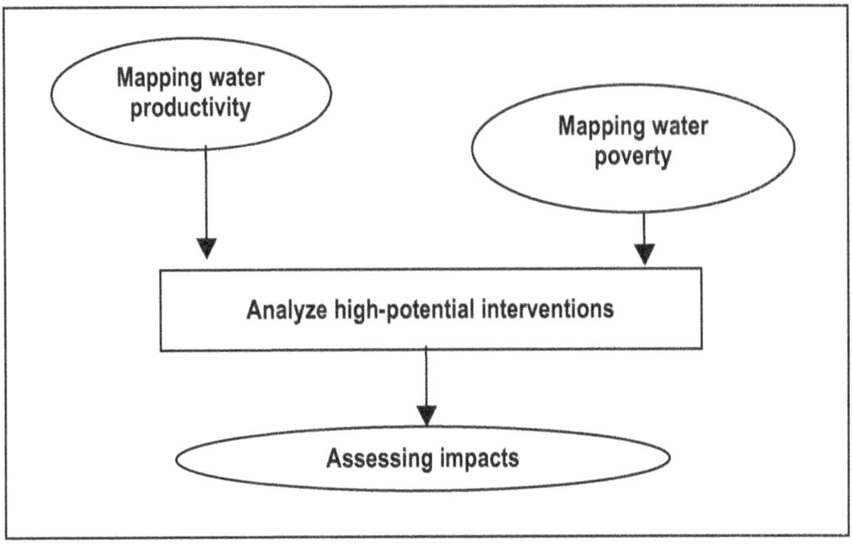

Figure 2.2 IWMI research: A new conceptual framework.

Each of these four themes has direct linkages to the new research framework. *Basin Water Management,* for example, *provides* the overarching context for IWMI's research on water productivity and water poverty across the hydrological cycle at the basin scale, and sets the agenda for the development, application and impact assessment of interventions at finer scales aimed at improving agricultural productivity, rural incomes, and human and environmental health. Within this context, *Land, Water and Livelihoods* focuses on identifying and testing technological, policy and institutional interventions to conserve resources and increase land and water productivity, while *Agriculture, Water and Cities* explores the rural-urban interface and interventions that can help ensure the safe and productive use of wastewaters and the sustainability of high input peri-urban systems. Finally, *Water Management and Environment* examines farm-, field-, and system-level interventions to better balance productivity and environmental objectives as well as the impacts of proposed interventions at the basin scale. Issues of policies, institutions and human health remain an integral part of each of these themes and to the overall IWMI research agenda. Thus, rather than compartmentalizing these issues, we have integrated them across IWMI's research portfolio as crosscutting 'Communities of Practice'.

Together, the new research framework and supporting themes aim to (a) increase the understanding of land and water productivity and its relationship to poverty, (b) identify promising interventions to improve the productivity and sustainability of the natural resource base as well as the access to productive resources, and (c) assess the impacts of such interventions on productivity, livelihoods, health and resource sustainability. In turn, we believe this sharpened research focus will help the Institute better carry out its mission to improve water and land management for food, livelihoods and nature, and, from that fundamentally contribute to the development of international public goods that support the CGIAR System Priorities and the achievement of the Millennium Development Goals.

3

Integrated Water Resources Management

Hammond Murray-Rust and Hugh Turral

3.1 INTRODUCTION

Historically, water management for agriculture has been equated with the development and operation of water systems and structures, largely for irrigation purposes. The Green Revolution technologies in particular relied on more irrigation water to allow food production to keep pace with global population growth and nutritional needs. The rapid development of water resources, however, has been accompanied by various well-documented costs such as negative environmental impacts, land degradation (principally salinization) and inequitable distribution of resource access amongst the rural poor. In addition, the rapid growth of urban centers and industry has led to increasing competition for water across sectors.

Future projections of population growth, changes in food preferences, climate change and urbanization all imply the need to increase food production through intensification ('more crop per drop') and expansion of irrigated area, where possible and environmentally acceptable. Irrigation is the keystone of current food security, but the options for further water development in all

 More Crop per Drop. Edited by M.A. Giordano, F.R. Rijsberman, R. Maria Saleth.
ISBN: 9781843391128. Published by IWA Publishing, London, UK.

regions except Africa are limited and competition for water is increasingly evident in Asia. It is clear that the pressures for reallocation of water from agriculture to 'higher-value' (municipal and industrial) uses are inevitable and, over time, will occur in more or less all countries. Thus, the key challenge now for agricultural water management is how to increase food production for a growing population while simultaneously meeting the water-quality and -quantity requirements of other economic and environmental sectors.

Meeting this challenge necessitates the evolution of strategies that reposition agriculture (a consumptive user of water) in relation to urban development and industry (generally non-consumptive users of water). It requires the provision of water resources for environmental needs, through changed policies on in-stream water diversion and off-stream storage, and a negotiation of the balance between food, livelihood and environmental priorities. Agriculture will have to produce more food with less water and this requires widespread improvement in the 'productivity' of water, both in irrigated and rain-fed farming systems, as well as in the reliability of agriculture in the face of increasing uncertainty and variability in water supplies from rainfall, streams and groundwater sources. Finally, addressing this challenge requires a holistic and integrated approach that considers the social, economic and environmental impacts of various solutions at a range of temporal and spatial scales.

Irrigation management has been at the core of IWMI's research agenda since the Institute's inception in 1984. Research in this area was initially focused on the operation, maintenance and efficiency of irrigation systems at the field and system scales. In response to the challenges outlined above, the thematic focus of IWMI's work underwent a sea change in the mid-1990s, when strategies to improve the productivity of water for food and livelihoods became the central focus. This restructuring of the theme, Integrated Water Management for Agriculture (IWMA), involved the development of new methodologies and tools to measure and evaluate agricultural water use and the accompanying trade-offs over time and space in terms of food production, environmental security and urban and industrial water needs. While management and operation of irrigation systems remain a key priority within the theme, the expanded scope of the theme has offered a more holistic and integrated assessment of water management in agriculture.

3.2 FROM IRRIGATION SYSTEM TO RIVER BASIN

IWMI Research Report No. 1 titled: *The New Era of Water Resources Management from 'Dry' to 'Wet' Water Savings* (Seckler 1996) represented a major change in IWMI's overall view of water management. Prior to the publication of this seminal document the focus of research was at the irrigation system level, as befitted the former name of the Institute, the International Irrigation Management Institute (IIMI). The transition from IIMI to IWMI was indeed much more than a cosmetic change in the Institute's name. It allowed the newly created IWMI the opportunity to place irrigation management into the overall context of river basins and to begin examining the interlinking hydrologic, socioeconomic and environmental aspects of water management at a variety of spatial and sectoral scales.

Classical studies of irrigation management had rarely looked at the issue of water at the basin level. Most studies looked at water management between the head of irrigation systems down to the point where water entered drains or went to deep groundwater and treated these fluxes as losses. Much of the focus was to improve deliveries to users within the system and reduce perceived losses by increasing efficiency, without really understanding what was happening to the water that left irrigation systems in terms of its reuse by others.

As described above, Seckler (1996) focused our attention on four implications of moving from irrigation system level to basin level (IWMI's basin paradigm):

(a) The importance of understanding the recycling of water within river basins.
(b) The importance of knowing whether basins are 'open' or 'closed'.
(c) The effect of scale on the interpretation and importance of water use efficiency.
(d) The need to look at longer-term trends in water supply and demand.

The concept of 'open' and 'closed' basins helped in determining which management strategies are most suitable. In open basins, where there are unused or unallocated flows out of the basin, improving efficiency is not necessarily the most effective strategy. There are options to increase supplies or transfer water out of a basin to where there is a greater need, or indeed to specify and reserve flows for the environment. Provided there are no adverse consequences in terms of waterlogging or salinization, unwanted water can flow back to drains or recharge groundwater, allowing reuse at a later stage. In closed basins all water is, by definition, already used for environmental or human requirements because there are no flows out of the basin and any reallocation becomes a trade-off between former and new benefits. In closed basins, the focus on improving water productivity becomes increasingly important.

Recycling of water within basins means that the classical concept of water use efficiency does not necessarily apply: for example, what was considered a loss in the upper part of a basin might actually be somebody else's water supply further downstream as a return flow via streams or groundwater. Such loss and reuse pathways can occur across the full length of a basin. This made us recognize that while individual irrigation systems may be inefficient in the classical sense, recycling may capture those losses, thereby ending up with high levels of basin efficiency. Indeed, closed systems are highly efficient because all water is depleted.

Prior to this, local-level water management activities rarely took into account either total basin water availability or the likely trends for each basin. Water management improvement programs were standardized whether or not water availability justified the approach adopted. Seckler urged us to look at overall water demand and supply at global and basin levels as a means to help us formulate appropriate and useful water management improvement strategies given the particular stage of water resources development in each basin. In order to link these ideas to IWMI's future research focus, Seckler challenged us to design our research activities that would help improve water productivity and proposed four interventions to accomplish this (Seckler 1996):

(a) Increasing the output per unit of applied (and transpired) water.
(b) Reducing losses of water to sinks and evaporation.
(c) Reducing the deterioration of water quality that inherently leads to reductions in potential productivity of that water when it is reused.
(d) Switching from lower-valued to higher-valued uses of water.

These actions require interventions at the field, irrigation system, catchment and basin scale. The four ideas became embodied into the revised research agenda of IWMI, particularly within the IWMA Theme, but by no means exclusive to it. The overall focus changed from improving water management from the traditional agronomic perspectives of higher yields and higher total production (land productivity) towards water productivity, where the focus is on improving the output from each unit of water used. IWMI's slogan therefore became 'more crop per drop'. In more recent times, it has become an essential part of the larger picture in the valuation of water and its use.

Seckler's 'basin paradigm' based on the four interventions listed earlier, has had a profound impact on IWMI's research thrusts and agenda, and the current structure of the IWMA Theme clearly reflects the research thrusts required to meet the challenge he laid down.

The first research priority, water productivity, is self-evident. It links IWMI to both our agronomic and economic colleagues in an effort to try to better define different meanings of water productivity, find ways in which water can be managed at all scales so as to enhance productivity, and turn these into practical tools for implementation. It is broader than IIMI's original mandate because it includes rain-fed water productivity, a necessary step if basin-level water productivity is to be improved.

The second research focus, integrated modeling, is structured primarily to address the issue of trade-offs between different water users not only within the agriculture sector but also between different sectors. To date, most of IWMI's efforts have been devoted to the application of GIS and remote sensing as tools to assist in integrated modeling. A major emphasis is on various forms of decision-support systems that allow policymakers and managers to assess the potential implications of alternative water-allocation strategies, based soundly on IIMI's pioneering work on irrigation performance indicators.

The third research focus, irrigation system management, is the continuation of IIMI's former mandate, recognizing that the largest fraction of all diverted water used by humans is for irrigated agriculture, and that irrigation and drainage system management needs constant improvement to meet food targets with increasingly stressed water supplies.

3.3 WATER PRODUCTIVITY: INDICATORS AND USE

The paradigm of 'more crop per drop' moved IWMI away from the traditional measures of water management performance based on agronomic principles, to those based on the economic principle of the maximization of output per unit of water. IWMI pioneered the change in thinking from yields per hectare to yields per cubic meter, and water productivity is now common in both scientific and popular writings. Measuring water productivity is now a standard when assessing water management performance, be it at field, system or basin level.

Despite the apparent simplicity of water productivity as a concept, in reality it has proven to be a far more elusive parameter than originally anticipated. In order to bring clarity to the research undertaken, IWMI has had to pursue two strands of work that when linked together give us a much clearer view of what productivity is, and how it can be improved through different management strategies: development of water productivity performance indicators and their application, and basin-scale characterization and resource assessment techniques.

3.3.1 Water productivity indicators

In its simplest form, water productivity is a ratio between crop output and water delivered. Conceptually this is not unlike the similar ratios used for irrigation efficiency, and can be applied at different levels in the same way that efficiency can be assessed at field, subsystem and system levels. Water productivity (Table 3.1) can be measured with respect to transpiration (important in closed basins), or water delivered at the field, farm-gate or at system level (including rainfall), but still needs to be related to land productivity, which up to now has been the main determinant of improved water productivity. Both physical production and its value have importance in different contexts of subsistence and livelihood, and cross enterprise comparison.

In most irrigation systems, and certainly at the basin level, output cannot be simply expressed in terms of kilograms because of diverse and complex cropping patterns. Only in those locations where there is monocropping, such as the rice-wheat systems of the Western Indo-Gangetic Basin, can production be used with confidence. In all other systems, some standardization has been necessary, and the most commonly adopted method has been some form of value. In a single system or country, value can be expressed in terms of local currencies. However, for broader comparisons where there are significant differences in local prices for the same crop, standardized gross value of production (SGVP) that takes into account yields, local prices for any crop, local prices for a base crop (typically wheat or rice) and world prices of the base crop is used.

Table 3.1 Water and land productivity indicators.

No	Indicator	Expression
1	Physical output per unit of cropped area	$(kg/ha) = \dfrac{\text{Production}}{\text{Area cropped under irrigation}}$
2	Value of output per unit of cropped area	$(\$/ha) = \dfrac{\text{Sale value of product}}{\text{Area cropped under irrigation}}$
3	Value of output per unit of irrigation supply	$(\$/m^3) = \dfrac{\text{Sale value of product}}{\text{Diverted irrigation supply}}$
4	Value of output per unit of water consumed	$(\$/m^3) = \dfrac{\text{Sale value of product}}{\text{Volume of water transpired by crop}}$

Source: Derived from Sakthivadivel *et al.* 1999b and Molden *et al.* 1998.

Water productivity indicators developed by IWMI (Molden *et al.* 1998; Molden *et al.* 2001a; Molden 1997; Sakthivadivel *et al.* 1999b) therefore address the relationship between available water at different locations within a basin. However, there is also a need to take into account the purpose for which the water was intended, and here the term 'process' is introduced. Process is not restricted to agriculture, applying equally to urban and domestic uses, but refers to the intended beneficial or productive use of that water. The suite of water productivity indicators therefore also includes the productivity of process water, and the productivity of water consumed at plant level through evapotranspiration. This provides the direct linkage with the agronomic and plant genetic interests of improving plant water productivity where output per unit of transpiration is the primary objective (kg/ha).

3.3.2 Water productivity: Case studies

With the establishment of robust water productivity indicators, IWMI was now in a position to undertake a series of case studies that would result in some benchmark values for water productivity for different crops in different conditions (see Table 3.2). A series of different case studies have been conducted, the most important being in Sri Lanka (Molden *et al.* 1998), India (Elkaduwa and Sakthivadivel 1999; Bastiaanssen *et al.* 1999a, b; Hussain *et al.* 2000; Hussain *et al.* 2003), Pakistan (Hussain *et al.* 2000; Tahir and Habib 2000), China (IWMI 2003), Turkey (Kite and Droogers 2000b; IWMI and GDRS 2000), Iran (IWMI 2004) and Central Asia (Murray-Rust *et al.* 2003). In analyzing the study results the main conclusions are the following:

(a) Water productivity values are relatively low in terms of income for most major grain crops, typically ranging from \$0.04 to 0.10/m^3 available when expressed in SGVP[1] terms. The implication of this is that grain crop cultivation is not likely to be an important pathway to reducing poverty. Their importance for food security is obvious, but they do not generate large incomes. Farmers who are confident that sufficient food will be available for purchase will likely opt for higher-value horticultural crops if market conditions are favorable.

(b) There is much more consistency in values for water productivity in terms of process consumption (actual evapotranspiration), generally in the range of \$0.10-0.14/m^3 of ETa, again expressed in terms of SGVP. This reflects the physiological process of crop growth and harvest index, the variations mostly being in terms of how much water is actually evaporated rather than transpired.

---

[1] The calculation of Standardized Gross Value of Production normalizes the value of different commodities in different countries in relation to purchasing power parity: the methodology is given in detail in IWMI Research Report 20 (Molden *et al.* 1998).

Table 3.2 Water productivity indicators for various systems around the world.

Country	System	Years	Water Productivity ($/m^3)	
			Output/unit of irrigation supply	Output/unit of actual water consumed
Burkina Faso	Gorg	1994/95	0.08	0.12
	Mogtedo	1994/95	0.11	0.15
	Savili	1994/95	0.28	0.62
Colombia	Coella	1993	0.14	0.20
	Saldana	1993	0.12	0.17
	Samaca	1993	0.63	0.34
Egypt	Nile Delta	1993/94	0.12	0.11
India	Mahi-Kadana	1995/96	0.07	0.06
Malaysia	Muda	1994/95	0.38	0.10
Mexico	Torreon	1996	0.12	0.20
	Alto Rio Lerma			
	Surface+Public wells	1994/95	0.18	0.24
	Private wells	1994/95	0.26	0.37
Morocco	Triffa Scheme Sec. 22	1994/95	0.27	0.34
Nepal	West Gandak	1996/97	0.13	0.12
	Khageri	1996/97	0.08	0.13
	Marchwar Lift	1996/97	0.36	0.12
Niger	Saga	1993/94	0.12	0.13
	Kourani Baria I	1994	0.05	0.17
	Kourani Baria II	1994	0.06	0.11
Pakistan	Chistian sub-div	1993/94	0.04	0.05
Sri Lanka	Nachchaduwa	1994/95	0.04	0.08
	Uda Walawe	1996/97	0.08	-
	Rajanganaya	1994/95	0.06	0.11
Turkey	Sarigol	1996	0.62	0.38
	Alasehir	1996	0.57	0.46
	Turgutlu	1996	0.30	0.46
	Manisa	1996	0.20	0.30
USA	Big Thompson	1996		0.13
	Imperial ID	1996	0.29	0.29
	Panoche WD	1996	0.37	0.38

Note: ID = Irrigation District; WD = Water District
Source: Sakthivadivel *et al.* 1999b.

(c) Water productivity for horticultural crops is significantly higher. In Turkey, where almost all crops studied were cash crops (grapes, cotton, fruits, vegetables), water productivity values were as much as $0.50/m^3. In Iran, vegetables can produce nearly $0.20/m^3, although inflated rice prices make rice cultivation equally profitable. Water productivity values increase with deficit irrigation[2] but are associated with significant decreases in total productivity and farm-level income. This makes deficit irrigation unattractive, as most farmers will not respond to high water productivities but to total farm production or higher farm incomes.

(d) Water productivity values decline when water quality is suboptimal. Salinity and sodicity of irrigation water depress already low values in marginal areas, enabling the real cost of land and water deterioration to be quantified (Kijne *et al.* 1998; Kijne and Kuper 1996).

(e) Low values of water productivity mean that there are few economic tools available, based on pricing of water that can be used to improve water use efficiency (Perry *et al.* 1997). It also means that grain production would be uneconomical if farmers have to directly compete for water with urban and industrial users at current commercial rates.

3.3.3 Water-saving technologies

IWMI has been involved in three major action-research activities that target the development and wider adoption of water-saving techniques at field and irrigation system level. In each study, all involving a range of national partners, IWMI's main interest has been to see the extent to which apparent savings at field level can translate into wider savings at system level. Conventional wisdom states that if only farmers would be more prudent in water use on their farms, then there would be widespread savings in water. Oweis *et al.* (1999) addressed some of these issues in relation to water harvesting and supplemental irrigation, but the analysis was largely restricted to field and small watershed levels, and did not really address the issue of basin-scale trade-offs in changing water use from one location to another. IWMI's results from the three studies show a more complicated picture:

---

[2] Deficit irrigation refers to suboptimal water supply (less than the full crop water requirement). Deficit irrigation can be beneficial for specific crops (stone fruits) and can stimulate increased yield or quality, if precisely managed. The classical example of deficit irrigation is in wheat, where one or two irrigation turns at tillering and flowering make a considerable increase in yield over rain-fed supplies in Northern India but fall well short of satisfying full crop water needs. This well-known fact formed the basis of warabandi—broad-based sharing of suboptimal water supply—in the 19th Century.

(a) The SWIM-supported 'more rice, less water' project in China highlighted the dilemma encountered when scaling up from field to irrigation system (Guerra *et al.* 1998; Matsuno *et al.* 2000). With careful control of water application at field level it is possible to reduce water application for rice by allowing fields to partially dry between irrigation turns. This reduces surface runoff, deep percolation and evaporation from the soil surface in the early growth stages. Thus, at field level, we find higher levels of water productivity because yields are not significantly different from traditional irrigation techniques.[3] But, at system level, we find little change in water productivity because there is widespread reuse of both surface runoff and water that has percolated into groundwater. These 'losses' turned out not to be losses to the system because of pumps and downstream reservoirs. While there may have been marginal gains due to reduced evaporation, these also were not captured at system level because their magnitude was too small to permit significant changes in discharges to irrigation subsystems. The dramatic change in allocation of reservoir water from agriculture to cities and industry has been compensated by extensive water harvesting by small-scale storage within the irrigated areas, which has been complemented by higher water productivity, though with lower overall production (due to reduced area).

(b) Similar results are obtained from work in the Rice-Wheat Consortium sites in Haryana, India and Punjab, Pakistan for which IWMI has been responsible. Gains in water productivity of up to 30% are made at field level due to adoption of resource conservation tillage methods such as land leveling, zero tilling and raised beds, and the importance of these benefits should not be underestimated. But at subsystem and system level, there are no significant changes in irrigation system operation, and thus we do not find the field-level benefits translating into water savings at system or basin levels (Tyagi *et al.* 2003).

At the heart of this is the difference between diverted and depleted water. If non-depleted but diverted water remains usable within the basin, it is not a loss, as it can be recycled. Thus, it is important to know where in the basin there is scope for 'real' water savings that can be effectively reallocated for other uses. This requires a better understanding of the spatial and temporal nature of water use and availability, or water accounting.

---

[3] There are interesting linkages with studies on malaria control (Van der Hoek *et al.* 2001) from the Water, Health and Environment Theme (see chapter 7), which suggest that wet-dry irrigation techniques may, under certain conditions, limit mosquito breeding.

3.4 WATER RESOURCES ASSESSMENT

While improvement in water productivity is necessary, it is not sufficient to me*et al*l water management challenges. What is required is a careful assessment of water resources on a spatial scale within the basins with a view to identifying distinctive zones where different water management strategies can yield 'real' water savings for additional allocation. For this purpose, IWMI research has developed a number of water accounting and basin characterization tools that allow the design of location-specific water management interventions that are best suited to the water and land resource potentials of basins and their segments. These tools can be applied both at the national and global levels.

3.4.1 Water-scarcity mapping

One IWMI product that has had a major impact is the water-scarcity map (Seckler *et al.* 1998a; Rijsberman 2000). This map gives a clear picture of locations where water is critically scarce, where it is scarce, and where water scarcity is unlikely to be encountered over the next couple of decades. It also indicates where scarcity is physical, due to lack of water, and where it is economic.[4] The concept of institutional water scarcity was also introduced, where neither water nor the economic means of development is limiting, but where the institutional and political situation prevents adequate and equitable provision of water for human needs.

Originally, the concept was applied using national statistics, resulting in a map (Figure 3.1) that gave an overall global picture of where water scarcity was most likely to be felt. However, for certain larger countries, notably India and China, which have significant variations in climate and hydrology, this was insufficient to guide policymakers. Where possible, therefore, the statistics have been disaggregated to basin level to give a much clearer indication of where basins are effectively closed or will become so in the near future.

Further tools for simulation have been developed in partnership with IFPRI, to undertake more detailed assessments, which also factor in trade in food commodities, such as IMPACT-Water (see Rosegrant *et al.* 2002) and its successor, WaterSIM.[5] This disaggregated approach links well with the PODIUM model that also has been adapted from the national to the basin level,

---

[4] Economic water scarcity means that while there may be additional available water, governments do not have the capital to develop them sufficiently to meet national water needs.

[5] WaterSIM or the World Agricultural Trade Simulation Modelling System, was developed jointly by IFPRI, IWMI and other partners. It has been recoded in GAMS at IWMI, and has been in operation since March 2004.

making it possible to view water management and food security from both the basin and national perspectives (Seckler *et al.* 1998a; de Fraiture *et al.* 2001; CA 2003). The Podium model was developed to allow an analysis of future water needs on the basis of scenarios of population growth and changing dietary preferences, which drive food requirements. The model was developed in spreadsheet format, with a custom-made interface for easy use. It has been distributed widely and used by many IWMI partners to understand future water needs.

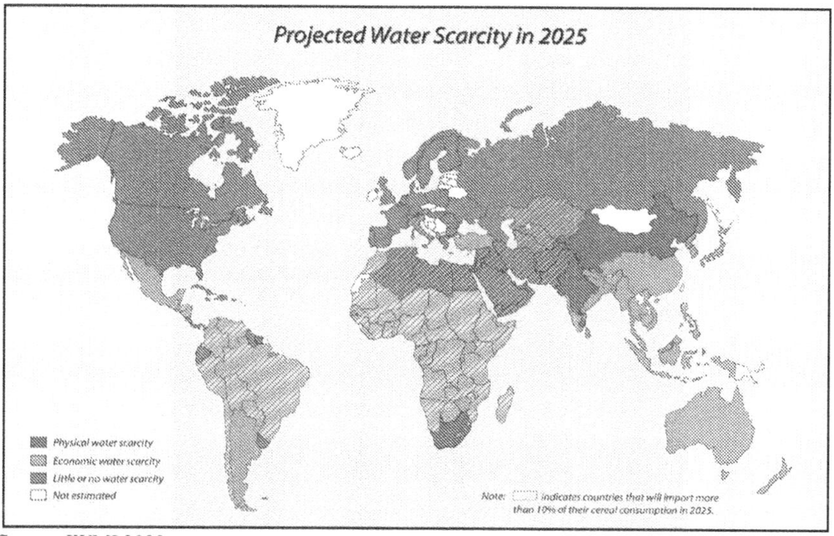

Source: IWMI 2000.

Figure 3.1 Predicted water scarcity by 2025.

3.4.2 Water accounting

IWMI pursued water accounting with the main focus on the utility of water and the extent to which it is used for beneficial purposes (Molden 1997). Water accounting can be applied at any scale from basin, subbasin or irrigation system, or even on individual farms. Several advantages come from this approach:

(a) Total water depletions are identified, and any surplus remaining is regarded as an outflow to the next level of analysis.
(b) Beneficial use of water is divided into two distinct components: process use, or its originally intended use at the point of diversion, and

non-process beneficial use where the water provides benefits that were not included in the initial allocation process.

(c) Estimation of low or non-beneficial depletions of water, notably evaporation or flows to sinks, provides one area of focus of changed water management that can free up non-productive water.

(d) Allowances are made for downstream commitments such as water rights or environment, the balance being potentially available, but only counted as available if flows can be stored or captured.

The main thrust of this approach is to show how much water is actually depleted, where and for what use, compared to that available and the portion diverted. This can be simply expressed for average or extreme cases by the 'finger' diagram (Figure 3.2), which separates depleted and non-depleted flow across beneficial and non-beneficial uses. The primary utility of water accounting is twofold. It permits the total amount of non-beneficial but depleted water to be estimated, which is essential for developing strategies to improve water productivity. It also helps clarify the net benefits of reallocation of water between different beneficial uses, and ensure that the main condition for moving water to its higher value is actually met.

There is a tendency to be too dogmatic in assuming that the initially intended beneficial use of water is automatically the most beneficial. Studies of total water productivity, which include both process and non-process use, indicate that in some circumstances the non-process uses may provide as much value as the process uses (Renault and Wallender 2000). The total value of livestock, aquaculture, agroforestry, orchards, vegetables and other crops not included in the original or official cropping pattern, is significant and may actually generate more cash income for farmers than food grains grown for consumption. Unfortunately, there are several biases that lead us to think in terms of food grains: most departments of irrigation and agriculture only deal with major crops, remote sensing normally only looks at major crops, and SGVP is most commonly used in conjunction with grains crops. Bakker *et al.* (1999) went further and identified a large number of uses of irrigation water that go far beyond the original purpose, and which have important health and social benefits that are not easy to quantify in financial terms.

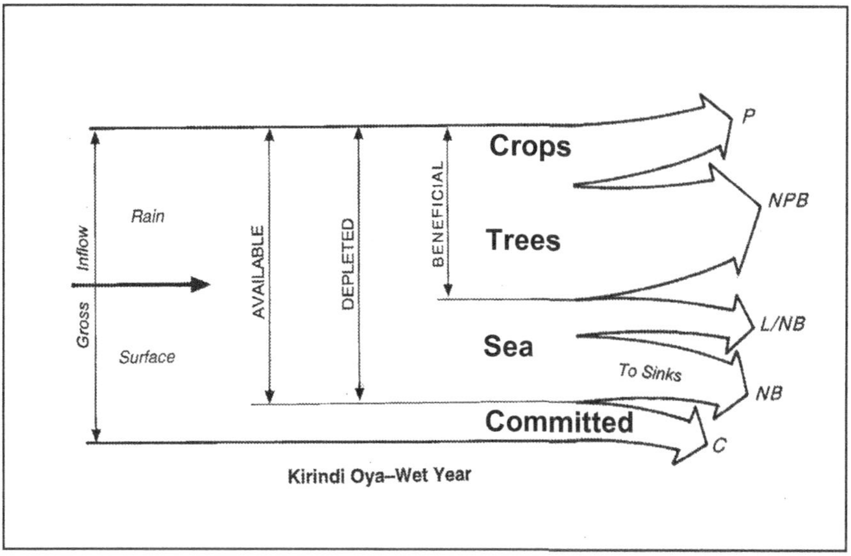

Notes: The key to use categories are: P = Process fraction (crop evapotranspiration); NPB = Non-Process Beneficial (forest evapotranspiration); L = utilizable/non beneficial (flow to sea/ beneficial environmental flow); NB = Non-beneficial (e.g. to saline groundwater); C = Committed. *Source*: Molden 1997.

Figure 3.2 IWMI's water accounting concept: The finger diagram.

Water accounting helps focus our attention on these non-process benefits, and helps us avoid the risk of reallocating water when there is no net benefit, or even potentially a loss in total benefits. Water accounting indicators lead us through a logical sequence of assessments: how much of gross water available is depleted, how much of available water is depleted, how much was depleted by the intended processes, and how much depleted by beneficial non-processes. This clarifies what types of management approach best suit the area under analysis (see Table 3.3). It appears to give clear and unambiguous strategies under a wide range of different conditions, as is evidenced by the case studies from Sri Lanka, Pakistan, and India in SWIM 1. Parallel work is now under way in South Africa.

Table 3.3 Water productivity: Scale considerations, process and indicators.

Scale	Crop	Field	Farm	Irrigation system	Basin
Processes	Water and nutrient uptake and use, photosynthesis, etc.	Tillage, fertilizer application, mulching	Distribution of water to fields, maximizing income	Distribution of water to farms, operation and maintenance (O&M), fees, drainage	Allocation across uses, regulation of pollution
Scientific interest	Breeders, plant physiologists	Soil scientists, crop scientists	Agricultural engineers, agricultural economists	Irrigation engineers, social scientists	Economists, hydrologists, engineers
Production units	kg	kg	kg, $	kg, $	$, value
Water units (m^3)	Transpiration	Transpiration, evaporation	Evapotranspiration, irrigation supply	Irrigation deliveries, depletion, available water	Available water

Source: CA 2003.

3.4.3 Hydronomic zones

Parallel to the development of the concept of water accounting was the development of the concept of hydronomic zones (Molden *et al.* 2001b). Water accounting is scale-neutral and independent of the topology of a given basin. While it indicates the types of changes in water allocation from one area to another, it does not indicate where management interventions should be focused.

Hydronomic zones enable us to characterize a basin into different types of water environments, each of which has a separate set of possible management strategies to deal with maximizing water productivity, water use efficiency and minimizing pollution or salinity. Of particular importance is the identification of the size and location of zones where improvement of classical water use efficiency is a legitimate strategy because outflows will only go to sinks or the sea and hence be lost. This strategy complements and needs to be combined with improvement of classical water use efficiency in zones where there is no recycling without threat of water loss, waterlogging or salinity.

The importance of this approach is that by introducing a spatial dimension to the assessment of different water management strategies, a more focused and more efficient method for basin-level water management strategies can be devised. This complements the second research focus of the IWMA Theme, the use of integrated modeling which is described in more detail below.

3.4.4 Trajectories of basin development

IWMI also recognizes that basin characterization efforts need to take into account the temporal changes in the level of basin development. Recent work by Molle (2003) distils an approach to characterizing basin trajectories and the technical, economic and sociopolitical influences that drive them. Molden looked at the relationships between total renewable water supply, potentially available water supply given current levels of technology, available water given current levels of water resources infrastructure and actually depleted water (Molden *et al.* 2001). In the earliest stages of water resources development in basins, actually depleted water is a relatively small fraction of potentially available supply and the focus of water resources development is likely to be on making more water available. As more water is depleted the focus changes to one of improved utilization within each water use sector, requiring more emphasis on management. When depletion levels are close to potentially available water supplies, there is a need to consider more stringent water allocation policies.

This approach links well with characterization of different institutional settings and IWMI's institutional and economic research. What is still required is a greater knowledge and use of the basin characterization tools and methods that can help policymakers better understand the consequences of different actions. This is highlighted in a case study from Nepal (Bhattarai *et al.* 2002) where the recommendations for the Indrawati Basin include a mix of improved technical understanding of water availability and utilization, a need for understanding the trade-offs between different water uses in the basin and selection of the most appropriate institutional mechanisms to assist in effective management of the total water resources of the basin in an integrated manner.

3.5 MODELING WATER MANAGEMENT

Once a basin starts to close, any change in one water use will impact on all other water users. Under these conditions it is not sufficient to adopt new management strategies that improve water productivity or crop productivity at a single location without also assessing the impact of these interventions on all other users in the basin. Impacts can be of two main types: effects on other water users within the agriculture sector involving issues of equity, water rights and other water allocation issues, and effects on nonagricultural users of water, notably on health and environmental issues. We increasingly understand that, with increasing urbanization and industrialization, the nature and form of irrigated agriculture will inevitably change as water is rerouted through cities, its quality (and therefore availability) is degraded and much increased volumes of wastewater are generated. In the North China Plains, average abstractions by agriculture have declined from more than 85% to between 60 and 65% in the last 20 years and indicate the potential for change. Clearly, these are areas with considerable interaction with all the other IWMI research themes (some of the issues of integration were addressed by Calder (1998) and Batchelor *et al.* (1998)).

For IWMI to grapple with this complex and difficult issue, significant resources were directed towards the development of models that could help us understand what the effects of changes from current water use might be. McKinney *et al.* (1999) assisted in this by reviewing a large number of approaches to modeling, although they tended to favor optimization models with a strong economic component. Models help us with two important concerns:

(a) To understand current processes so that we can simulate current conditions. These are often physical processes, such as hydrology or soil-plant-water relationships, but they also include decision-making processes, operating rules and other institutional matters.
(b) To be able to predict the impact of possible changes in resource availability, resource quality or decision making through some form of scenario analysis. To this effect we need a range of performance indicators that can assess impacts from the perspective of all water users in the basin, not just for agriculture, so that we capture the interdependency of water users under increasing water scarcity.

In this regard, IWMI has also made some important choices that guided model-based research. These decisions are:

(a) Wherever possible, models should be in the public domain so that our clients would not be faced with potentially expensive purchases of model software.
(b) IWMI itself would not develop new models but would work with existing models and, if necessary, make modifications to better suit our needs.
(c) Where clients already use models, we should not try to make them change, but try to use these models in conjunction with other products, which IWMI feels would help them better understand water management.

(d) We need to have clear linkages between models at different scales, from field to basin, so that all water users would be addressed in the scenario analysis phase.

With this philosophy in mind, IWMI has developed an approach to modeling that now enables us to look with considerable confidence at the interaction between different water users within a basin, and has reached a point where effective model-based decision support systems can be developed for specific basins.

IWMI has also devoted much effort to the development of *tools* that provide data for modeling exercises, using 'new' technologies such as GIS and remote sensing. A prime focus of integrated modeling efforts has been to understand basin-level water accounting and the nature of water productivity at different scales within the basin. IWMI and its partners have invested heavily in the development of remote sensing applications in agricultural water management, resulting in a state-of-the-art book by Bastiaansen (1998).

3.5.1 MODELING WATER ALLOCATION

The spatial aspect of water management has also been addressed through IWMI's work in conjunction with the Stockholm Environmental Institute with the modification of the simple WEAP water balance and allocation model to better incorporate agricultural concerns. The model has been used in scenario analysis of water allocation under global climate change scenarios in California (Huber-Lee *et al.* 2003) and has potential for broader application to basins with limited data.

A somewhat different approach has been tested in South Africa where efforts have been made to try to understand the relationships between different water users in the Steelport River Basin (Stimie *et al.* 2001). This links directly to the Theme 4 research on institutional issues related to basin management, but includes an assessment of both the demands for each individual water user and the interrelationships between water depleting and polluting upstream users and their counterparts downstream who are directly impacted by reduced water flows and water quality.

3.5.2 Modeling crop water productivity

Results from case studies provide benchmark figures for actual practices but they do not really tell us much about the potential water productivity, because it is impossible to control all of the variables. Results from one location may not be achievable elsewhere due to differences in climate, soils, crop genotypes and local cultivation and irrigation application practices. Multifactor determinants of water productivity have been determined from farm-survey data using multiple linear regressions, which often show that fertilizer level has a greater significance than water application in Pakistan (Hussain *et al.* 2003).

IWMI therefore decided to try to use simulation models to generate results that complement data derived from field studies (Ines *et al.* 2001; Kite and Droogers 2000a; ADAPT 2004; Aerts 2004). The majority of these results have been generated from the SWAP model, with some contributions from DSSAT.

The advantage of simulation modeling is that, once a model is calibrated, it can be used to assess the impact of variations in key input parameters on yield, water productivity and water balance. Typical input parameters that can be varied include total water application including rainfall, individual irrigation applications and their scheduling, and water quality in terms of salinity and sodicity. Crop modeling has a long history, but mainly in research, development and validation of models, with less concern about their application in water resources allocation and policy settings. It has only recently been applied to answer questions that arise at higher spatial scales than the field or research plot. The main outputs from the use of simulation modeling at crop level are the following:

(a) To estimate yields and water productivity for any combination of water availability and water quality for each crop-soil combination. These can be expressed in the form of simple quadratic equations that allow quick and simple calculation of key performance indicators. The modeling approach appears robust and matches closely with field-data collection activities. Given that field studies are time-consuming and, therefore, relatively expensive, the use of well-tested simulation models is an effective strategy for continued work.

(b) To determine the maximum attainable water productivity for each location, and therefore to identify the gap between actual data from field studies and maximum probable output. For example, in the Zayandeh Rud Basin in Iran, average basin-level water productivity for agriculture is currently \$0.12/m^3 of water available for agriculture for existing cropping patterns (including rice, tree crops, horticulture and fodders). Assuming the same cropping pattern but an adoption of a wide range of improved water management practices at farm level (drip irrigation, pipe conveyance systems, precise application of irrigation water, land leveling, etc.) the absolute maximum basin-level water productivity that can be obtained is \$0.18/ m^3 of available water, a 50% increase, and will probably be somewhat less. In addition, these benefits require substantial investments and additional annual costs, so that net margins will be far lower. Any further increase in water productivity can only be obtained by switching to higher-value crops.

(c) To distinguish between E and T in the ET term as long as it can give good simulation of leaf area indexes (LAIs). This is of particular importance because a major strategy in improvement of water productivity is to sustain high levels of transpiration while minimizing evaporation from the soil surface. The SWAP model results suggest that for most crops there is a sharp decline in evapotranspiration as soon as LAI reaches unity, so that water savings in terms of evapotranspiration can generally only be made during crop establishment phases. While E for rice may be as much as 400 mm/season out of a total of some 1,000 mm for ET, for most field crops E is only 150-200 mm out of a total ET of 800 mm.

3.5.3 Modeling with remote sensing data

Assessing water productivity at broader scales that cover irrigation systems and whole basins presents special challenges. Case studies are expensive and time-consuming, particularly with respect to water measurement at system and basin levels, while modeling runs the risk that calibrations may be inaccurate when scaled up to large areas. To avoid both of these situations, IWMI has invested a great deal of effort in researching the potential application of remote sensing and other techniques that can help assess water productivity over wide areas.

Remote sensing techniques offer opportunities to estimate a number of parameters that can be used to help improve water management.

When IWMI started researching the possible applications of remote sensing for water management only two satellite platforms were widely available: expensive, relatively infrequent but high resolution Landsat 5 images and free but coarse resolution AVHRR images from NOAA. During the past decade more satellite platforms have become available that make remote sensing more realistic in terms of both cost and resolution. Perhaps the most useful for our work are the 250 to 1,000 m resolution, daily, free images of MODIS that provide a wide range of different spectral bands and multiple processed products, such as EVI (enhanced vegetation index).

Using remote sensing, IWMI has pioneered three main applications for water management. They are described below:

First, basin-level water productivity and land cover assessment have been successfully used in several different locations, including Turkey, Pakistan, Iran, Central Asia, India and Sri Lanka. Although NDVI has been widely used as a way of determining land cover, IWMI has been able to link these data with those of other satellite sensors estimating sensible heat by using both SEBAL and SHEBA procedures to estimate the evaporative fraction and hence actual evapotranspiration.

(a) The combined outputs of SEBAL, SHEBA and other more standard measures provide us with the first real opportunities to determine moisture stress (the ratio of actual to potential evapotranspiration) and estimate biomass growth and crop yields. This means that using remote sensing it is possible to calculate several significant performance parameters over wide areas that complement data derived from field-level studies (Kite and Droogers 2000a).
(b) In both Pakistan and Turkey, basin-wide assessments of yields and water productivity were determined for the first time. In Pakistan, due to the arid climate, the results were limited to the irrigated areas of the Indus Basin, while in Turkey they were extended to include estimated water productivity and yields from rain-fed, groundwater-irrigated and surface-irrigated areas (Droogers and Kite 2001 a, b).
(c) The Turkey results again support the findings of other studies that water productivity in terms of value per cubic meter are higher for rain-fed and supplementally irrigated areas than for fully irrigated areas, but the total production is lower (Kite and Droogers 2000b).
(d) Results of field studies in India and Pakistan indicate the severity of impact of salinity and sodicity on water productivity. With increasing values of EC and SAR, water productivity declines sharply, and it is possible to use remote sensing to better understand the importance of location within canal systems on water productivity and yields.

Second, irrigated area mapping, using several different parameters including NDVI and SEBAL analyses, has been a major thrust to develop practical applications of remotely sensed data (Bastiaanssen 1998; Bastiaanssen *et al.* 1999b). Few countries have reliable measurements of their actual irrigated area, often reporting instead the potential or equipped area. However, except for arid areas, estimating irrigated areas from satellite images has proven elusive. Rain-fed areas in the wet season will generate similar reflectances to irrigated areas, while dry-season irrigated areas may not extend over the whole area irrigated in the wet season. By using multiyear analyses and with limited ground truth, IWMI has been able to develop techniques, which provide estimates of irrigated area, which are superior to other sources of information.

Third, estimation of potential water productivity at basin and national levels has been conducted through a combination of remotely sensed data, the IWMI's

Climate Atlas[6] and other secondary information (Droogers *et al.* 2001). For the Indian subcontinent, monthly moisture availability indexes provide guides to optimal planting dates and the length of the reliable growing season where moisture is the predominant limiting factor. For southern Africa similar techniques were used to assess rain-fed potential.

IWMI's first experience with a proper integration of modeling and remote sensing (Kite and Droogers 2000a) was in the Gediz Basin in western Turkey, a basin that is closed in the growing season and where competition for water between different uses is becoming quite severe. During this project IWMI learned a number of lessons on how integrated modeling can shape future research on water management. Among these lessons were:

(a) No single model is able to address the linkages between field, irrigation system and basin scales because different scales of analysis require different types and details of data. For some aspects, such as soil-plant-water relationships, models may need detailed and specific data, while at basin level it may be possible to use more generalized information.
(b) Every basin is likely to have data deficiencies that require a flexible approach to the modeling process. If data are simply not available, then it is necessary to select an alternative model that is not dependent on that set of information even if there may be some loss of accuracy involved. Some data are particularly elusive, especially those relating to decision rules for the operation of reservoirs and other hydraulic control structures, and actual discharges into irrigation systems. Field-data collection and interviews remain indispensable.
(c) There needs to be increased clarity of what performance parameters are required to be included in the analysis, particularly those that deal with environmental and social impacts of water management, so that interaction between different uses of water can be included in the analysis. Without these additional performance parameters, the results will likely be biased towards hydrological and agronomic impacts, and ignore other concerns from other sectors. In the Gediz, for example, we successfully examined the trade-offs between agricultural production and the requirements of wetlands with importance for bird life (De Voogt *et al.* 2000), but in the absence of satisfactory indictors of what the actual ecological requirements were, we had to make a great many assumptions.
(d) Considerable use can be made of public domain information, particularly from the Internet, to supplement or even replace more

---

[6] Available at: http://www.iwmi.cgiar.org/WAtlas/atlas.htm

traditional secondary data sources, which may be missing, inaccurate or hard to access because of bureaucratic constraints. Of particular value are Internet sites with climatic information, soil and land cover information, discharge and flow records, and reservoir data.

(e) Remote sensing and the use of GIS are indispensable. Public domain remote sensing data provide the basis for accurate land cover maps, assist in delimiting irrigated areas and other land uses, provide estimates of actual evapotranspiration and form the basis for GIS-based databases. In the Gediz, it transpired that national statistics on irrigated areas were significantly under-reported because farms using privately owned tube wells were not included in the data. Without the more accurate information provided through remote sensing, integrated modeling would not have been possible.

(f) No satisfactory models exist for assessing the hydrology of irrigation systems, which is far more important in this type of research than understanding the hydraulics of irrigation systems. The Turkey experience made us recognize the importance of assessing the magnitude and location of return flows and groundwater recharge, although quantification was difficult.

The Gediz study was successful in linking basin-level hydrologic models with field-level evaluations of water productivity and related performance parameters. It was possible to determine yields, production, water productivity at basin, system and field levels under a range of different water availability conditions (Kite *et al.* 2001).

The study was also successful in developing a satisfactory framework for scenario analysis that dealt with both exogenous and endogenous factors. Endogenous factors such as climatic change and land cover change are beyond the control of managers within a basin but require response if and when they occur. Endogenous changes, such as water-allocation strategies or changed field-level water management practices by water users, are the basis for most decision-support systems.

The integrated modeling approach proved transferable to the adjacent Kucuk Menderes Basin using only public domain data derived from the Internet, and it was able to assess the viability of a proposed new reservoir in that basin (Lacroix *et al.* 2000).

Two other partial modeling studies were carried out that have similar to the components of the Gediz study. A hydrological model of the Mekong Basin was developed along more or less the same lines as was done in Turkey, relying heavily on Internet data and remote sensing (Kite 2000). The focus of this study was to simulate the impact of upstream water resource developments on the viability of fisheries, particularly those in Cambodia and the Mekong Delta, in collaboration with ICLARM (now known as the WorldFish Center). The results of this study are greatly restricted by lack of data on existing water-control infrastructures and their operating rules. This highlights the fact that integrated modeling is not merely an extension of a hydrologic model but that it involves a far wider range of different types of data. A similar situation was experienced in southern Sri Lanka, examining interactions between irrigation system management and the hydrology of coastal lagoons (Renault and Makin 1999; Hussain *et al.* 2000) where we also ran into serious data problems.

Based on the Turkish experience, a second major effort using integrated modeling was undertaken in the Zayandeh Rud Basin in Central Iran (IWMI 2004). This is a severely stressed basin that has been effectively closed for 40 years, but where demand for water continues to increase. There are also severe and increasing threats from salinity and sodicity, an immediate risk of groundwater mining, and a desire to sustain rural incomes and livelihoods even though water allocations for agriculture will inevitably decline.

The overall approach to modeling and the integration of remote sensing and GIS in the Zayandeh Rud is given in Figure 3.3 and the way that modeling at different scales is integrated is shown in Figure 3.4.

The main improvements in the Zayandeh Rud study over the Gediz study were the following:

(a) Considerable effort was made to compile as comprehensive a database as possible from a variety of different sources, including both technical information and the management and operational targets of a wide range of different agencies and institutions.

(b) Development of a salt and water balance model using a simple spreadsheet that could accurately predict salt accumulation and river salinity under different hydrologic conditions. Because the Zayandeh Rud is completely regulated upstream of any major abstractions, a complex hydrological model was not needed, being replaced by an empirical model of water releases and allocations between sectors and between irrigation systems.

(c) The use of specific groundwater models to identify the source of groundwater and assess the extent of groundwater mining, and investigate long-term trends in groundwater.

(d) Incorporation of information on downstream water-quality deterioration due to agriculture, and urban and industrial sectors to determine minimum flow requirements.

(e) The use of water requirement projections for the next 20 years based on available information including possible trans-basin diversions into and out of the basin.

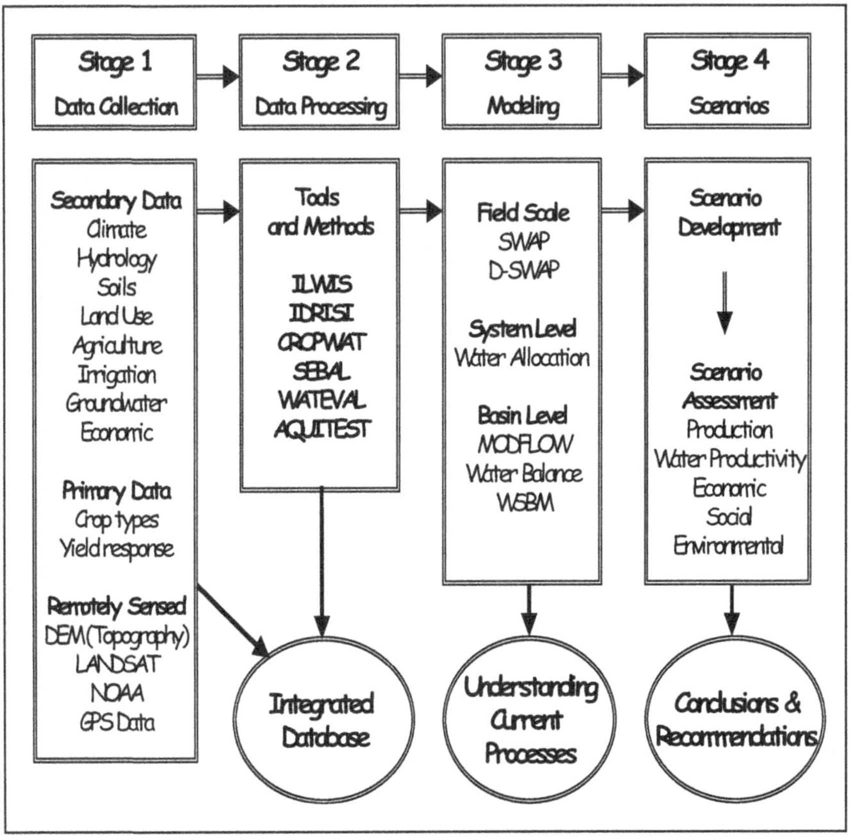

Source: Murray Rust *et al.* 2005.

Figure 3.3 Schematic diagram of integrated modeling approach in Zayandeh Rud, Iran.

(f) Development of a more wide-reaching scenario assessment framework that had three important developments:

(i) Identification of six different water availability conditions ranging from highly stressed to above normal that were used in all of the different models to ensure consistency in output from the different models.

(ii) Identification of a range of management scenarios at both irrigation system and field levels that cover the range of practical options.

(iii) Incorporation of a much wider range of performance indicators at basin, system and field levels that addressed economic, environmental and equity concerns.

(g) Development of an integrating spreadsheet that used outputs from each of the different models and calculated the values of all of the performance parameters identified as being relevant to basin needs and conditions. This proved to be far more effective than trying to simulate all conditions in the basin in one model.

(h) Significantly improved capacity to look not only at changes in the value of different performance indicators but also examine trade-offs between them. This is particularly valuable when there are issues such as equity, incomes and environmental concerns competing with more traditional objectives such as yields and water productivity.

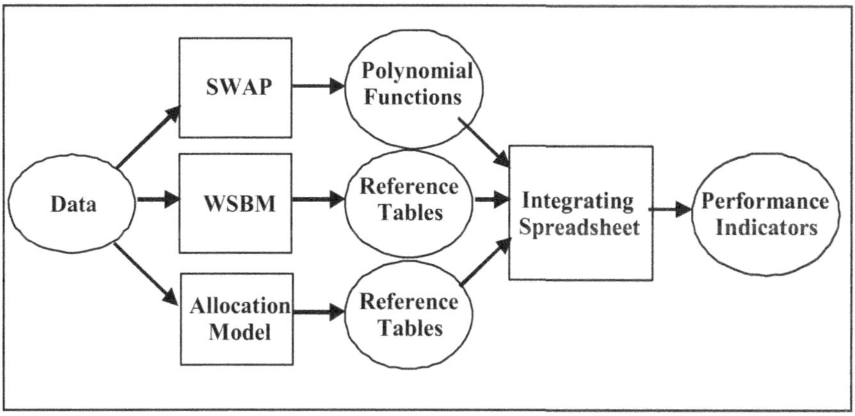

Source: Murray Rust *et al.* 2005.

Figure 3.4 Integration of modeling results at different scales in Zayandeh Rud, Iran.

Given that both the Gediz and Zayandeh Rud Basins are water-stressed, it is not surprising that similar conclusions were drawn from the research but many of the conclusions in the latter study go beyond those of the Turkey case study. The breadth of these research results is the direct benefit of an integrated modeling approach. Without the use of models, plus associated database and remote sensing resources, it would not be possible to come up with specific results and recommendations for the Zayandeh Rud. The process of integrating modeling is replicable and, currently, significant progress is being made to establish similar integrated models in the Krishna Basin in India and in IWMI's own benchmark basins in South Africa and Pakistan. However, the models and their applications are still evolving, and work in this area needs to continue into the future. From IWMI's perspective, the focus still needs to be on the impact of changed water allocations and management practices on the agriculture sector, and particularly on its ability to provide sufficient food to meet demand under a global scenario of shrinking water resources for agriculture.

The integration and relationships between different models, working at different scales, has been neatly summarized by Droogers (pers. comm. 2004, Wageningen) and is shown in Figure 3.5, which relates physical detail (on an inverted scale) to the scale of investigation, and shows where IWMI has a selection of models that deal with surface water and groundwater problems or at global scale with prediction of food and water demand PODIUM).

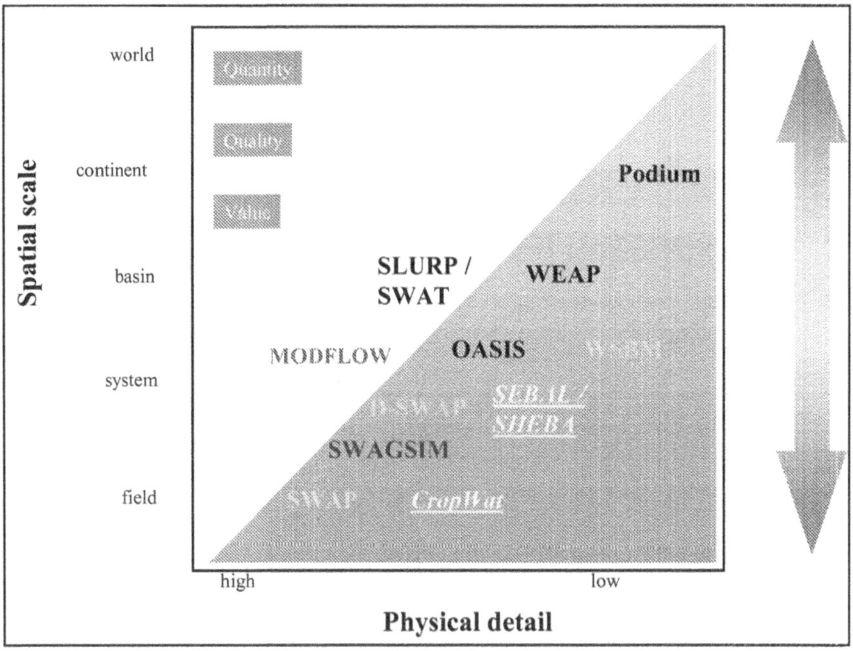

Figure 3.5 Models with different spatial scales and physical details.

3.6 IRRIGATION SYSTEM MANAGEMENT

Despite the large shift in emphasis towards water productivity and integrated modeling at IWMI, research still continued in the former core subject of irrigation management. However, there have been changes in the emphasis in response to three factors:

(a) The general decline in funding for irrigation management which has reduced the opportunities for undertaking research on operation, maintenance, rehabilitation and irrigation performance, which had been central to IIMI's research.
(b) An increased concern with management transfer and the consequences of transfer on how irrigation systems continue to function following transfer.
(c) A greater concern about institutional aspects of management, as opposed to technical issues because there is concern that if management capacity is weak then operational and design changes become less important.

Despite these changes, IWMI has accumulated an impressive set of publications and research findings related to different aspects of irrigation system management during the period covered by this review.

3.6.1 System O&M

Almost all of the research undertaken by IWMI on irrigation O&M focuses on the issue of level of service. This is in considerable contrast to the emphasis in the 1980s on the rational approach to irrigation scheduling that had a strong agronomic imperative supported by concerns with canal hydraulics. The level of service concept accepts that users have a major stake in how water is allocated and delivered, and that operation should be structured so that users are satisfied with the performance levels. This concept applies equally to both more traditional large-scale systems where government agencies control water in the main and secondary parts of the system and systems that have experienced total or partial transfer.

The vast majority of IWMI's results indicate that both the conceptualization and implementation of service-oriented irrigation management remain extremely weak. In a few cases, notably in transferred systems in Turkey, water deliveries are user-driven and highly predictable, essentially removing water as an uncertainty for irrigating farmers (Murray-Rust and Svendsen 2001). But in most cases, rules and regulations remain unclear and in almost all studies the conclusion is the same: without an improved level of service, it is improbable that irrigators will be able to maximize their farm incomes and improve water productivity.

Indeed, in many cases, uncertainty of surface water supplies encourages a switch to either partial or even total dependence on private pumps. While this relieves some of the uncertainty at individual level, it makes system management much more complicated, decreases its revenue stream from water fees and greatly undermines the willingness of water users to work together in a cooperative mode for system operation, management and improvement.

The single common recommendation is that individual and community water rights have to be clearly defined and established before operational planning will be acceptable to all water users (Perry 1996; Perry and Narayanamurthy 1998; Albinson and Perry 2002). Water rights have to be clear at all levels from field to basin; it needs to be completely clear as to whether they are volumetric or a proportion of available water; and there must be clear rules about how to cope with droughts. This requires dialog with users to determine what is considered fair in terms of water rights, something that is still generally lacking in systems, irrespective of whether some degree of management transfer has occurred or not.

The research also generally concludes that there are no technical fixes that can overcome the lack of adoption of better service orientation (Bandaragoda 1998b). While some results suggest that structured irrigation systems, where rigid design and allocation principles are followed, make water users more efficient in their water use, evidence from case studies suggests that a few structured systems really work as intended, and that levels of service remain low (Brewer *et al.* 1997; Murray-Rust *et al.* 2000; van Etten *et al.* 2002).

Research in two locations on system-level water allocation decisions shows rather contrasting responses to unexpected shortages in water availability. In the Sri Lanka case study (Sakthivadivel *et al.* 2001), when it became clear that there would be insufficient water to irrigate the entire area, dialog between the government agency and water users resulted in an acceptable water-sharing agreement that enabled most farmers to obtain some benefit from limited water. However, the success of this single-season experiment was limited because no long-term lessons were learned to extend and strengthen the dialog process, and exclude politicians from making arbitrary interventions into water allocation decisions at system level. It appears that the stress on system managers to cope with the severe shortage of water could not be maintained when the water situation became somewhat better in subsequent seasons.

In Turkey, major operational changes were implemented in the Gediz Basin following an extended drought (IWMI 2000). The operational rule changes which included precise scheduling of irrigation deliveries with full discharge for limited time periods were accepted by water users and have been continued following the drought because they resulted in a significantly higher level of service. This was a positive example of learning from adversity rather than merely muddling through and reverting to normalcy after the crisis period.

Despite a great deal of work on irrigation system operation, there is little evidence that managerial solutions are properly linked to the design and purpose of the systems (Renault and Godaliyadda 1999). More site-specific solutions that are well grounded in principles of level of service remain to be fully developed.

3.6.2 Rehabilitation and modernization

A not dissimilar set of conclusions come from IWMI's limited work on rehabilitation and modernization. All our research took place where there was some form of dialog with farmers and accompanying efforts to rehabilitate or modernize irrigation systems.

The studies arrived at a clear understanding that users have to be involved in dialog before design changes are contemplated. The Gal Oya System in Sri Lanka, the first major effort to involve water users in design and construction of rehabilitation projects as well as in subsequent O&M, shows positive benefits from a combination of technical and institutional innovations, with improvement in a wide range of performance indicators (Amerasinghe *et al.* 1998). Similar conclusions come from studying interventions in small tank cascades (Sakthivadivel *et al.* 1997).

However, in the North-West Frontier Province (NWFP) and Sindh in Pakistan where there was little or no involvement of water users in the rehabilitation process, design changes were largely ignored because they were viewed as abstract or alien by water users (Renault and Makin 1999; Lashari *et al.* 2000; Lashari and Murray-Rust 2000). As a result, performance improvements in these systems were negligible, if not adverse, because lack of attention to water rights and existing use of water meant that inequity between head and tail areas often increased as a result of poorly conceived interventions.

While methodologies for improved dialog and consultation with water users have been mooted and pilot-tested in a few locations in Sri Lanka, India and Pakistan, the reality is that where rehabilitation still proceeds it has not sufficiently learned the lessons from previous projects. As a result, these interventions remain costly and ineffective.

3.6.3 Performance of irrigated agriculture

Much of IIMI's interest in performance assessment has focused on the improvement of management of irrigation systems through a systematic process of target setting, monitoring and evaluation, and target revision in light of actual performance. IWMI has broadened this approach to look more at performance indicators that measure the impact of irrigation management through indicators that permit greater comparison across systems. The two approaches are complementary and target different audiences. The original set focuses on irrigation managers as they are internal to irrigation systems and concern the process of management, whereas the second set is for policymakers and planners external to individual systems and deals largely with output.

The comparative indicators include a mixture of output-oriented indicators that cover both the production and productivity and the water supply and financial indicators (Molden *et al.* 1998). These have been tested in a range of different types of systems in different agro-ecological regions. In Mexico, efforts were made to help irrigation districts use performance indicators to improve their management (Kloezen *et al.* 1997; Kloezen and Garcés-Restrepo 1998). While some success was noted, the indicators do not always meet the social and institutional goals of managers and users alike. However, the principle of developing and using performance indicators was seen as highly beneficial. Turned-over systems in Turkey use their own performance indicators rather than those developed by IWMI, but the principle of having annually reported targets and accomplishments remains firmly in the forefront of the Irrigation Association practices (Svendsen and Murray-Rust 2001). In South Africa, indicators are included that directly assess the performance of management in addition to output and impact indicators (Tren and Schur 2000). Several studies in India have used the same approach (Sakthivadivel and Gulati 1997; Sakthivadivel *et al.* 1995a, b; Sakthivadivel *et al.* 1998).

The utility of using performance indicators for improving management can be seen in the increasing adoption of benchmarking. IWMI, in conjunction with ICID, IPTRID and the World Bank, has established international norms for indicators on the free access Benchmarking website,[7] and managers and policymakers from many countries can use this facility to compare their performance with similar systems in other places. A significant aspect of this development is that people increasingly accept that, while the principle of setting targets and seeing the extent to which they are met is a universal procedure, it is up to individual managers to set the value of the targets themselves.

[7] The site address is: http://www.lk.iwmi.org:82/oibs/LoadBench.htm

Performance monitoring remains an arduous task insofar as it requires a great deal of monitoring. IWMI has experimented with the use of GIS as a cost-effective method for assessing performance in the Bhakra Irrigation System in India. Although the methodology appears to give good results, it remains a technique in its infancy and requires an effective supporting database before it can be systematically adopted (Sakthivadivel *et al.* 1999a; Sakthivadivel *et al.* 2001). Nevertheless, there have been concerns that performance still remains too focused on output and economic productivity and not enough on other less-tangible social and environmental issues.

The concept of multifunctionality of irrigation systems includes such diverse benefits as flood protection, recharge of groundwater, disposal of wastewater, landscape protection, erosion control, and biodiversity and environmental protection. In Taiwan, these functions are afforded high value by the public in general, meaning that agricultural production alone is an insufficient basis for evaluating the performance of irrigation systems (Matsuno *et al.* 2002).

Overall, IWMI and its forerunner IIMI have made significant contributions to the development of performance indicators and their application for managers and policymakers. The extent, to which they are adopted, however, remains a largely institutional matter, and it is not surprising to see increased linkages between the IWMA Theme focus on indicator development and the WRIP Theme concerns with their adoption by management organizations.

3.7 CONCLUSIONS

In terms of impact on the scientific community, the IWMA Theme has made major steps over the years, particularly with respect to conceptual development and methodologies. IWMI has contributed significantly to the knowledge base through its literature on water management over a wide range of issues in all sorts of agro-ecological zones. The main advances in conceptual terms are related to water productivity, basin characterization and linkages between field, irrigation system and basin:

(a) IWMI had a global impact with the world water scarcity map and the distinction between physical and economic water scarcity.
(b) IWMI pioneered much of the thinking about water productivity and its importance at different scales and the formal literature still reflects this.
(c) The linkage among field-, irrigation-system- and basin-level efficiencies has been explored so that the goal of improving classical water use efficiency has been placed into a broader perspective of water resources utilization efficiency.

(d) There have been similar impacts with respect to basin characterization, particularly in terms of the dynamics of basin water resources development and the importance of changing demand and supply conditions on management needs and foci.
(e) There was a logical progression from the conceptualization of performance of irrigation systems to performance assessment at the basin level.

The success of these conceptual advances can be seen by looking at scientific literature. The ideas are widely quoted and have been absorbed into the main stream of thinking all over the world. It is no mean feat to accomplish this within a decade. Further, and no less impressive, are the methodological advances that have been developed to parallel conceptual developments, an approach that makes IWMI a practical rather than an academic institution.

Methodological advancements have also been made in several areas, and in many respects have really only been possible because of the conceptual advances. The most important of these have been the following:

(a) IWMI has led the way in developing indicators of water productivity and continued to refine and broaden performance indicators at both irrigation system and basin levels. Having promoted the concept of 'more crop per drop', IWMI provided the tools to measure this, and extend it to 'more value per drop'.
(b) Tools for basin characterization, particularly water accounting and hydronomic zoning, have broken new ground in helping us better understand what current and future water supply and demand conditions will be, and what the most appropriate management responses should be in each specific set of conditions.
(c) IWMI and partners have pioneered the use of remote sensing as a practical management tool, so that it is cheaper and quicker to undertake performance assessment studies than it ever has been. Being able to measure actual water use is a major step forward. Using more and more public domain data sets to minimize arduous field-data collection is a major methodological breakthrough.
(d) The use of individual models has been impressive in better understanding a wide range of relationships at the field and basin levels, particularly for soil-plant-water issues, water allocation, basin-level water and salt balance, but rather less so at irrigation-system level, where modeling of system hydrology linked to both field and basin hydrology remains an elusive goal.

(e) Integration of different models with different purposes within the context of a single basin has been developed and through that IWMI has fostered greater capacity to look at trade-offs between water, food and the environment.
(f) Decision-support tools are evolving and we can expect this to be refined and completed in the next research phase.

In both areas of conceptualization and methodology, IWMI has been more involved as a partner in the related activities. This is because the field of interest is broader, there are more players and there are needs for a greater range of skills and methodologies than IWMI can afford to provide through its own staff. Therefore, IWMI shares credit with other institutions and organizations much more than it used to, which we view as a sign of increasing collaboration rather than reduced influence.

Overall, looking at the challenge set by Seckler (see section 3.2) helps us to place the IWMA Theme accomplishments in perspective. We have made good progress in our research related to interventions 1 and 2 from conceptual and methodological perspectives, but our impacts are still modest. We have not made much progress with respect to intervention 3, "reducing the deterioration of water quality that inherently leads to reductions in potential productivity of that water when it is reused". In particular, we have not capitalized on much of the conceptual work laid down by IIMI with respect to salinity and sodicity. Intervention 4, "switching from lower valued to higher valued uses of water", is more an issue of markets, and the IWMA Theme is not well equipped to handle that. However, through the combined efforts of the IWMA and WRIP Themes, IWMI could offer a unique perspective on water use decisions, not in terms of case studies or assessments, but on how hydrologic and economic incentives can be enhanced to encourage farmers to switch to higher-valued uses of water.

In light of these accomplishments as well as the remaining unfinished business, we have reframed our IWMA research priorities. Under a new title, Basin Water Management, we focus on completing the challenges posed by Seckler (1996) to improve the productivity of water. As described in more detail in the concluding chapter, the objective of the new Theme is to provide a better understanding of the trade-offs and options in agricultural water management at the basin scale and contribute to improved equity and productivity in water use through the development of appropriate tools and methodologies for analysis and management. Understanding water productivity at the basin scale remains a primary focus of the theme, with complementary research on sustainable water use in agriculture and institutions policies, and economic instruments for better water management at the basin scale.

4

Smallholder Land and Water Management

Frits Penning de Vries and Deborah Bossio

4.1 INTRODUCTION

Under its research theme 'Smallholder Land and Water Management', IWMI tries to promote integrated thinking in land and water resources management and effective actions with the application of science to the practical level for smallholder agriculture. Although it became a formal theme only with the Strategic Plan 2000-05, the research focus on smallholder issues is as old as IWMI, especially within the broad context of research on irrigation management transfer as well as in the specific context of research on water productivity and livelihoods. Research focus in this crucial area became more focused, however, as IWMI's mandate was broadened from 'irrigation' to 'water' beginning in the mid-1990s and with the merger of the International Board for Soil Research and Management (IBSRAM) with IWMI in 2001. With a broader organizational mandate and an inherited research experience and expertise, IWMI research under this theme has become strategically more focused on improving water and land productivity in rain-fed and fragile regions, promoting catchment conservation and reversing land degradation and its on- and off-site effects.

 More Crop per Drop. Edited by M.A. Giordano, F.R. Rijsberman, R. Maria Saleth.
ISBN: 9781843391128. Published by IWA Publishing, London, UK.

4.2 RELEVANCE AND JUSTIFICATION

Hundreds of millions of smallholder farmers live in the rural uplands and rain-fed areas of Asia, Africa, and Central and Latin America. Only 20% of the world's agricultural land is irrigated, the rest is 'rain-fed'[1] being dependent on rain that falls on that land only. In Sub-Saharan Africa, rain-fed agriculture accounts for 95% of the agricultural land and supports as much as 70% of the rural population. Ongoing poverty and hunger characterize the situation for 800 million rural residents of these areas who earn less than the equivalent of US$1 per day, and are classified by the World Bank as living in extreme poverty. They are chronically hungry, lack access to safe drinking water and are unable to pay for adequate health care or education for their children. Productivity per unit of land and per unit of water on rain-fed smallholder farms is generally far below its climatic potential.

In Sub-Saharan Africa, for instance, cereal yields in rain-fed systems are only one ton per hectare, far below yield potential, as compared to about five tons or more per hectare on irrigated lands. The causes of low productivity are myriad, and include poverty, highly variable rainfall, lack of access to inputs and markets, decreasing labor availability, and progressive degradation of land and water resources. In fact, the rural poor live disproportionately on the poorest lands (Table 4.1) and their situation is often complicated by the fact that women are overburdened with running their households as well as farming, and in Africa, HIV/AIDS has left unskilled orphans to farm for themselves.

Despite the difficulties, agriculture will remain the main source of livelihood support and engine of growth for these families (UN Millennium Task Force on Hunger 2005). Many smallholders can develop their natural and other resources into small agricultural enterprises, while others can provide inputs and services to the farmers or buy farm products and add further value for sale in urban markets. Indeed, it is widely thought that agricultural growth will continue to be the key driver of rural poverty reduction in the near future (Hazell and Johnson 2002), regardless of the trends in off-farm migration and transition to more industrialized economies. Improving the food security and income for these millions of smallholder farming households demands special attention in areas with historically low land and water productivity. Crucial are those who live in rain-fed systems without supplementary irrigation, who have not yet benefited from decades of public investments in infrastructure, services, and agricultural research and extension. Our challenge, therefore, is to identify and promote

[1] We use the term "rain-fed" in the broadest sense to include all agricultural systems that lie outside conventional surface water irrigation schemes; thus it includes a whole host of practices including pure rain-fed, a variety of rainwater harvesting techniques and supplemental irrigation.

ways in which smallholders can improve their productivity and income through improved management of water and land resources. Improvements must be viable under difficult external conditions, and achievable in a way that protects the natural resource base upon which farmers depend.

Table 4.1 Land degradation and location of the rural poor.

Region	Rural poor on favored lands (millions)	Rural poor on marginal lands (millions)	Rural poor on marginal lands %
Sub-Saharan Africa	65	175	73%
Asia	219	374	63%
Central and South America	24	47	66%
West Asia and North Africa	11	35	76%
Total	319	613	66%

Sources: Nelson *et al.* 1997; Scherr 1999.

The rationale for research to increase environmentally sustainable productivity in smallholder farming systems in upland and rain-fed areas emerges essentially from the indispensable need to address poverty and food insecurity in rural and fragile areas of Asia and Africa. The recent stagnation in the growth of agricultural production even in the Green Revolution areas has also prompted policymakers to look more toward rain-fed agriculture as a means to maintain the momentum of productivity growth that will be necessary to feed the growing world population. The negative impacts that unsustainable upland farming practices have on downstream populations and resources, and on biodiversity preservation, constitute an additional driver for this area of research that has lent increasing urgency to the task. It is for this reason that the *UN Millennium Task Force on Hunger 2005* and the *Inter Academy Report on potential of agricultural growth in Africa (IAC 2004)* give top priority to a range of specific actions to assist smallholders and their land and water management. More importantly, smallholder research is very critical for addressing, at least, three of the Millennium Development Goals (MDGs)[2] established by the United

[2] These are: MDG 1: Eradicate extreme hunger and poverty. Reduce by 2015 the number of people living on less than $1.00 per day to half the 1990 level, and halve the proportion of people who suffer from hunger in the same period; MDG 3: Ensure gender equality and promote empowerment of women. Women have an enormous impact on the well-being of families and societies—yet their potential is not realized because of discriminatory social norms, incentives and legal institutions; and MDG 7: Ensure environmental sustainability. The environment provides goods and services that sustain human development so we must ensure that economic development sustains the

Nations and accepted as yardsticks of development by most countries around the world. Clearly, IWMI research mandate under this theme fits directly with the current realities and global consensus.

4.3 RESEARCH CONTEXT AND BACKGROUND

Despite the poor state of upland and rain-fed areas in the world at present, there are tremendous opportunities to sustainably increase the productivity of smallholder systems to improve livelihoods for the rural poor in spite of, or taking advantage of, their challenging situation. Numerous examples of successful intensification of smallholder systems confirm these opportunities are real and achievable within local contexts (Pretty *et al.* 2004; Pretty and Hine 2004). Improved fertility management and supplemental irrigation, for example, can significantly reduce uncertainty, and overcome the chronic low productivity and crop failure that are characteristic of rain-fed agriculture (Rockström and Falkenmark 2000; Rockström *et al.* 2003; (UN Millennium Task Force on Hunger 2005). The recent Copenhagen Consensus results ("Putting the World to Rights", *Economist* 2004)[3] also confirm that significant opportunities exist for enhancing resource productivity, and with it food security and poverty reduction, through improved land and water management practices[4] and access to low-cost technologies.[5]

The Green Revolution in Asia stands as one potential development pathway for rural areas. In the past three decades, hundreds of millions of Asian farmers 'graduated' from the category of subsistence farmers as their farms became more productive; farm products could provide food security for their families and an excess to sell, and many smallholders left their farms and found off-farm employment. This Green Revolution primarily resulted from the breeding of cereal crops that responded well to intensive management, fertilizers and irrigation that could be supplied in lowlands. The result was that yields of crops such as rice and wheat doubled in less than 30 years (Plucknett 1999). The Green Revolution occurred in many of Asia's lowlands and terraced slopes where irrigation systems were already present and land user rights were well

---

environment. Better natural resources management increases the income and nutrition of poor people.

[3] The recent Copenhagen Consensus ranked small-scale water technology for livelihoods and water productivity for food production as two of the nine top investment opportunities to advance global welfare, particularly for developing countries.

[4] For example, low or zero-till agriculture, supplemental irrigation, groundwater recharge and water harvesting systems.

[5] For example, low-cost small electric and diesel pumps, manual devices such as treadle pumps, low-cost bucket and drip lines.

established. Extensive support by governments, through development of markets for inputs and outputs as well as through extension services, has facilitated the success.

This major success in large regions of many Asian countries is in stark contrast to the ongoing poverty and hunger in African and Asian upland and rain-fed areas, where the history of development efforts and of research is one of sporadic and limited success. The Green Revolution that is now criticized for substantial negative environmental consequences is also criticized for this inequitable distribution of benefits.

To address these gaps in success, research efforts in general have substantially changed over the last few decades. In particular, technology development and adaptation research has moved from an exclusive productivity and experiment station focus, to on-farm research. This builds direct partnerships with farmers and includes their active participation, allowing better integration of political, social and economic constraints of the farm families. Research on integrated natural resources management (INRM) has also evolved to link productivity research with environmentally sound management of natural resources (Javier and Voss 2003). Another positive trend in smallholder research is the move toward more holistic approaches with due attention to socioeconomic limitations, social development and long-term environmental impacts as exemplified in the livelihoods approach (Campbell *et al.* 2002). IWMI's research, under the theme of Smallholder Land and Water Management, has contributed to, and was built upon, these holistic INRM approaches by giving a particular focus on areas and contexts where water is an entry point for improved livelihood and environmental security.

4.4 SMALLHOLDER RESEARCH: EVOLUTION AT IWMI

IWMI's expertise in land and water management for smallholders is built on IWMI's early work focused primarily on productivity, equity and management of small-scale irrigation systems in Asia, and the integration of the programs of the International Board for Soil Research and Management (IBSRAM) into IWMI in 2001. This integration broadened the smallholder focus considerably by bringing issues of soil and land management firmly into IWMI's research agenda. Key features that IBSRAM brought to IWMI were: knowledge of soils and how farmers manage soils; participatory on-farm research in Asia and Africa; capacity building in networks; and research-extension linkages. This complemented IWMI's expertise in irrigation management and policy and institutional aspects. It also expanded IWMI's perspective to a broader view of the spectrum of land and water management options available to farmers - large scale, small scale, irrigated, rain-fed - to enhance food production and livelihoods, as well as integrating formerly fragmented smallholder research.

4.4.1 IBSRAM's experience with smallholder research

Since IBSRAM programs were merged with those of IWMI in 2001, its research and network experiences have provided a solid foundation for IWMI research on smallholder land and water issues. Created in 1985 and based in Bangkok, Thailand, IBSRAM was dedicated to assist the applications of soil science and promote sustainable food production in developing countries (Craswell 1998) in partnership with national agricultural research and extension systems (NARES) and other international organizations. A network or consortium mode of operation was used partly to overcome the problem of fragmentation arising from the location specificity of research on many land management issues and partly to promote a partnership approach that focuses on the needs of NARES and builds their capacity for research.

There was a continuous evolution in the research approach of IBSRAM with notable methodological and operational contributions. As collaborative activities evolved from a strong soil and on-station focus (IBSRAM 1988) to a broader sustainable farming focus (IBSRAM 1997), the research became more multidisciplinary and farmer participatory in nature, and new concepts for sustainable land management were explored. The organization contributed to the development of concepts such as participatory research (Rhoades 2001) and integrated natural resources management (INRM) (Sayer and Campbell 2002). Holistic and multidisciplinary approaches were involved in many IBSRAM programs. Instances in this respect include (a) eco-regional research, (b) the CGIAR system-wide program through the Soil, Water and Nutrient Management (SWNM) consortium, (c) the Framework for the Evaluation of Sustainable Land Management, and (d) the Resource Management Domains concept (see Smyth and Dumanski 1993; Syers and Bouma 1998; IBSRAM 1999; Coughlan and Lefroy 2001; Dumanski and Craswell 1998). IBSRAM was also involved in the development of the livelihoods approach (Campbell *et al.* 2002), with the inclusion of off-site effects and economics (Enters 1998; Drechsel and Gyiele 2000). This evolution in scientific research brought a new multidisciplinary perspective both in effectively combining knowledge and approaches from different scientific disciplines as well as in actively exploring how scientific knowledge can serve smallholder communities in actual practice.

IBSRAM coordinated networks as tools for capacity building and research. Its networks were collaborative programs of three to seven national partners that lasted up to 12 years. Networks did not attempt to solve problems directly but worked with national teams to help them solve specific soil-conservation and land use problems. Two of the main networks (*ASIALAND* Management of Sloping Lands and Management of Soil Erosion Consortium [MSEC]) were on erosion control in Southeast Asia and the results are reported below under Catchment Management (Penning de Vries 2002; Maglinao and Leslie 2001); two (*ASIALAND* Management of Acid Soils and *AFRICALAND* Management of Acid Soils) addressed soil-fertility and soil management issues on marginal land with acid soils and carried out field research on soil organic matter turnover and use of phosphorus fertilizer, as well as promotion of suitable laboratory techniques (Lefroy *et al.* 2000) and one (*PACIFICLAND* Management of Sloping Lands) focused on erosion control under annual and perennial crops in the Pacific Islands (Dowling 1996).

Two others focused on specific problem soils, the Vertisol Network and the Upland Soils Network (Craswell 1998). IBSRAM had a strong program to disseminate information on sustainable land management. Over a 10-year period, in Africa for example, IBSRAM conducted about 20 annual meetings and training workshops in 10 different countries. More than 300 soil scientists from up to 20 African countries took part. There were in addition six specialized conferences, e.g., on soil and water management of vertisols, as well as a data-quality management program for several NARS and countries. IBSRAM published a series of Technical Notes, with guidelines and manuals for field and laboratory research, whicht later became a Global Tool Kit Series;[6] a global information service via the quarterly: *Soil Management Abstracts*; and a regular newsletter.

4.4.2 Lessons from the network approach

In addition to scientific knowledge, summarized in the following sections, there were also interesting lessons learned from the procedures and impacts of the research and capacity-building networks. Highlighted here are a few of these useful, albeit subjective and qualitative lessons.

(a) The equal, nontraditional partnership between all partners was appreciated by partner organizations. Long-term cooperation can have a positive effect on the level and direction of the research of the partner organization (Maglinao 1998; Phommasack *et al.* 2002).

[6] These IBSRAM Tool Kit Series include: Tools for the Economic Analysis and Evaluation of On-farm Trials (1999) , Financial Assessment of Land Management Alternatives, Practical Guidelines for Data Collection and Calculation of Costs and Benefits (2000), IBSRAM Training Manual on Participatory Research and Technology Development for Sustainable Land Management (2000), and Socioeconomic Diagnosis for the Evaluation of Sustainable Land Management (2000).

(b) Joint research can be a cost-effective way to obtain and analyze more primary data and from a wider range of sites than would be possible without national partners (see the subtheme on 'Catchment Management'). It takes time and effort to integrate network concerns with partner priorities and this is not always successful.
(c) While the capacity amongst partners varies significantly from country to country, all researchers in the networks have benefited from cross-country exchange, and their perspectives have gained a higher profile within the international scientific community. Furthermore, networks help in promoting the ownership of research conclusions among all partners, and hence accelerate their uptake and implementation in partner countries.
(d) Steering Committees are essential for ownership of the network by all partners. Ample attention must be given to make them really effective, particularly in a dynamic social or scientific environment. Research issues and approaches are likely to evolve during the lifetime of a network. This may require new expertise and partnerships. Steering Committees tend to be conservative and may resist change. Involving extension services in research networks needs ample attention at the early stage and is ineffective as a late add-on. Without funding they are not sustainable.
(e) For MSEC, the network of instrumented catchments represents a significant resource for research that can be utilized to tackle a variety of issues as land use and societal values evolve; it is unique in the area.

Over time, research issues in the networks became more multidisciplinary and farmer-participatory in nature. Consequently, farmers and farmer organizations became partners in the social and biophysical research. They provided important feedback with respect to the relevance and efficiency of proposed activities, while their indigenous knowledge and insights were a rich source of information (Lefroy *et al.* 2000). Farmers also expanded the capacity for field research during participatory on-farm research where both farmers and researchers learnt. Drechsel and Gyiele 1998 and Drechsel 2000 provide a detailed review of lessons derived from these on-farm and network activities.

4.5 RESEARCH ACHIEVEMENTS

The focal point of IWMI's Theme 'Smallholder Land and Water Management' was how to improve water and land productivity of rain-fed and small-scale systems at the catchment scale. The research was divided in the subthemes: 'Productivity of Smallholders', which focuses on the adoption and adaptation of improved practices and technologies as well as on the equity issues of increasing the access to, and productivity of, water in smallholder agriculture; 'Catchment Management', which concentrates on INRM at the watershed scale to cover both on- and off-site impacts of resource management; and 'Rehabilitation of Degraded Lands' because degradation of natural resources is so important, and because land and water degradation processes are interlinked.

4.5.1 Productivity of smallholders

The Strategic Plan 2000-05 (IWMI 2001a) identified the subtheme 'Productivity of Smallholders' as an area of particular interest as it provides a nexus between increasing food productivity to alleviate hunger and poverty (first recommendation of the UN Millennium Task Force on Hunger. 2005) and the emerging major opportunities to achieve significant impacts with water and land that were recently supported by the Copenhagen Consensus (*Economist* 2004). IWMI's focus in this area is increasing productivity on smallholder farms via water management.

The three core research issues are: (a) increasing water productivity (for crops, livestock or other purposes); (b) improving access to water; and (c) integrated farm-/community-level multiple users and uses of water. These core issues are to be addressed within the biophysical and socioeconomic environment in which these issues emerge and interact. These issues are also to be considered in the wider policy context of upscaling practices and technologies from single farms to landscapes and service delivery (such as supplying treadle pumps) and benefit generation (such as clean water for downstream communities). Research reviewed here focuses on the core research issues noted above and how they relate to external biophysical and socioeconomic environments. Future work will continue to focus on integrating smallholder technologies into the broader context of livelihoods.

4.5.1.1 Improving water productivity of smallholder farmers

To increase income often involves using water and land resources more efficiently (increasing water productivity), through crop selection, reducing the intra- and inter-seasonal variability of the production process (i.e., irrigation to minimize drought impacts), and improving fertility management. These options for improved water and land management represent real opportunities to close yield gaps by increasing water use efficiency and productivity.

A key issue is efficient use of water in the crop production process itself. Water use efficiency is the outcome of different and competing processes. Water use efficiency on rain-fed smallholder farms is typically only 3,000 kg/ha of biomass in cereal crops (of which, only 1,000 kg are grain) for 800 mm of rainfall in a cropping season, or 0.37 kg/m^3. In comparison, irrigated systems' water use efficiency for cereal crops is often much higher, varying between 0.6 and 1.7 kg/m^3 for wheat or rice, and between 1.1 and 2.7 kg/m^3 for maize (Zwart and Bastiaanssen 2004). Two important avenues to increase water productivity on the farm are to reduce evaporative losses of water and to remove other constraints to growth such as nutrients, thus allowing water to be used more productively.

Effective traditional techniques to reduce evaporation, such as mulching, conservation farming, watering of plants through half-buried clay pots, and multistorey relay cropping, are well known and were widely practiced in the past. However, due to land and water degradation and the increasing cost of labor, they are employed less frequently and skills and knowledge about them are being lost. At the same time, new techniques, such as drip and sprinkler irrigation, plastic mulches, and even micro-greenhouses are now available that are more effective, reduce labor demands, and/or are less expensive.

Due to the wide range of information already available on these systems, research at IWMI has been limited in scope, and focused on understanding the mechanisms contributing to the efficacy of new techniques in high- and low-quality water situations, such as intensive small-scale water management and management of saline water. Based on field research in South Africa, Sally *et al.* (2000) confirmed that precision irrigation, such as drip irrigation, makes two to three times more water available for transpiration and plant production for upland crops. In areas with slightly saline groundwater, such as those in Sub-Saharan Africa, drip irrigation is better than sprinkler irrigation because it adds less salt to the soil and is less of a threat to sustainable farming (Karlberg and Penning de Vries 2004).

Soils on many smallholder farms are either inherently low in fertility, or have experienced nutrient depletion, and this limits the productivity of water significantly. To promote soil fertility, it is important to understand the driving forces behind nutrient depletion in smallholder systems. Research on fertilizer use on farms in Northeast Thailand (Konboon *et al.* 2001) was revealing in this regard. Nutrient depletion over time was common, but the large variation in fertilizer inputs between farms in the same district was also notable. Much of the variation between farms appeared to be related to differences in the proportions of off-farm and (non-rice) on-farm income. As the proportions of income from rice increased, so the productivity and the sustainability of the farms decreased. It was concluded that a lack of income from off-farm sources limited farmers' ability to purchase agricultural inputs. These results demonstrated that a multifaceted livelihood system was important not only to sustain the small farm enterprise, but also to prevent land degradation. In addition, all soils are not equal, in their original fertility levels, their vulnerability to degradation, or in the management that best suits sustainable production. Therefore, management of specific soils received attention in IBSRAM networks, which reviewed methodologies and practices for farm productivity and sustainability on problem soils in Africa, such as vertisols and acid upland soils (Craswell 1998).

4.5.1.2 Improving water access for smallholder farmers

Acquiring more water for farming has traditionally been practiced by diversion from streams, by harvesting water from other plots and leading it to cropped fields or storage facilities and by drawing from wells and surface water manually or by animal power. Such widely known techniques have made agriculture possible even in dry areas. However, land degradation, water degradation, sedimentation in river courses, and loss of social coherence in communities tend to reduce the use of some of these technologies (Oweis *et al.* 1999).

Recent improvements in manufacturing technology have brought new options for individual water management by smallholders. This development means that more individuals can afford implements to manage water without recourse to larger community- or government-owned equipment or facilities. Understanding the diversity and impact of different systems on-farm is a first step for planning of upscaling and extension. In response to this need, research to catalog and review very small-scale (individual scale) irrigation technologies has been a focus at IWMI in partnership with IPTRID.

A variety of traditional small-scale water supply systems have been extensively reviewed by IWMI and others (e.g., Mati forthcoming; Badiger 2003; Reij and Steeds 2003). More quantitative analyses focused on the treadle pump. The International Development Enterprise (IDE) has sold and facilitated the installation of over a million treadle pumps in South Asia (Shah *et al.* 2001). One pump, on average, generates $40-400 per year per household and allows a full return on all investments in one year or in a single cropping season (SIMI 2003; Badiger 2003). In situations where water is available and markets for produce exist, smallholder farmers, even those who are poor, are willing to invest their own funds to intensify vegetable production (Penning de Vries *et al.* 2002a; Van Koppen 2002).

The crucial implication of such a finding is that the financial sources for small-scale investments go far beyond those available from public sources. ApproTec, IDE and Enterprise Works obtained similar results in Africa, albeit, at smaller scales. Based on this knowledge of the technology and its potential, IWMI has helped design and promote a treadle pump for the South African environment (Seago 2003). Engine-driven water pumps have also been miniaturized and are becoming more affordable for affluent farmers, sometimes superseding the treadle pump.

Still, all the requirements are not really known for the major upscaling of the adoption of small-scale water supply and irrigation technologies among the poorest farmers and disadvantaged persons, including some women and HIV/AIDS victims. IWMI has recently started research on the processes of initial adoption and continued use of treadle pumps in Africa. Gender composition and labor cost are the two important features that have received attention. Results suggest that simple technologies can empower women with higher output relative to that of men. For instance, pumps can improve women's access to water, especially by reducing the need to transport heavy cans of water. A comparative analysis of irrigation technologies in West Africa showed that introduction of pumps increased both the employment of women on vegetable plots and their ownership of these plots (Drechsel 2003, pers. comm.).

4.5.1.3 Multiple water use: Efficiency and equity roles

Lack of women's rights to land and water for cropping and for household uses often leads to heavier workloads for women and lower production of the total system. For instance, in the command areas of new irrigation systems in Burkina Faso, land allocation policies rarely allowed women to obtain an irrigated plot, despite the fact that productivity of land and labor is higher in irrigation systems in which both men and women have plots (Zwarteveen 1997). The implication is that higher positive production and social benefits could result if smaller plots were allocated separately to men and women. These findings are important because women's income is generally more beneficial for dependents than that of men. Van Koppen (2002) states that a powerful indicator for performance of the irrigation system is the degree to which women farmers have access to land, water, other inputs and credit.

In rural and peri-urban areas, water is used for various domestic purposes. In many contexts, however, water is used or can be used for a variety of productive and income-earning activities such as gardening, field crops, animal husbandry and brick making. These uses may complement or be in competition with each other. Yet, in most cases, water sources, uses and users are not well integrated, leaving much scope for improvements in water use efficiency, productivity and equity. Examples of such improvements are more accessible, reliable and cleaner water for household uses and productive purposes. Means for achieving these aims are primarily new institutions that enable effective interactions between end users and water providers. Such improvements may increase the ability of water users to pay for the construction and maintenance of the systems and minimize water-sharing conflicts among users and uses. This has a crucial role in creating the necessary conditions for an effective management and accelerated upscaling of multiple-use systems (Moriarty *et al.* 2004). Recognizing this important role, IWMI has started research with support from the CPWF and a range of partners, including NGOs, on multiple uses of water in communities with particular focus on the disadvantaged and poor sections of communities such as women, subsistence farmers and HIV/AIDS victims.

4.5.1.4 Effects of the socioeconomic and biophysical environments

The response of the smallholders depends clearly on their biophysical and socioeconomic environment. To understand environmental driving forces, IWMI, in collaboration with several NGOs, evaluated the performance of indigenous water management technologies in different biophysical and socioeconomic environments in India and Nepal. Some of the centuries-old coping strategies and practices adopted by farmers with smallholdings and in fragile environments are not well known in scientific circles. With support from DFID, Badiger (2003) looked at the positive and negative aspects of six 'old' and 'new' systems that address all three-core issues of smallhoder research within the wider context of the biophysical and socioeconomic environment (in brief, all techniques listed in Box 4.1 were appreciated by users as 'effective', though for quite different reasons, Badiger 2003, but they were abandoned, once the biophysical or socioeconomic conditions were no longer favorable).

This holistic study resulted in a number of important findings related to efficacy, livelihood implications and social and economic aspects of the adoption dynamics for these innovative water management practices. It was found that all the technologies presented in this study are promising low-cost innovations relevant to smallholders in improving the availability and management of water for rural livelihoods. There are important climatic and physical factors determining the feasibility of the innovations that can be defined in a quantitative and useful way. With respect to adoption, three aspects were found to be important: (a) key players, i.e., adoption was greater with the presence of an NGO or progressive farmer who leads the process of initiation, transfer and spread of these innovations; (b) community composition, i.e., adoption of innovations was positively influenced by the homogeneity of communities and negatively influenced by income diversification; and (c) livelihood strategies, i.e., farmers who were primarily dependent on agriculture-based occupations were more willing to adopt, try or improve the innovation.

Box 4.1 Smallholder technologies in India and Nepal.

- *Paal:* Water-retaining structure in the 400-600 mm rainfall zone to force runoff to infiltrate into the soil, increasing the amount of water available for cultivation and to recharge groundwater. As a result, farmers were able to take up crops such as onion, millet, mustard and wheat during winter and vegetables during summer, and women became more involved in watershed management.
- *5% technology:* Runoff is harvested and stored in a pit about 5% of the size of the irrigated fields, which is particularly appropriate for upland agriculture and when rainfall is erratic. The technique helped reduce weather-related risks that affect upland paddy cultivation, and allowed vegetable cultivation and fish farming.
- *Integrated land and water management:* Communities make arrangements for land use (grazing, forest products) and water use (domestic, irrigation, groundwater recharge) specific to different parts of the catchment. Gradual improvement in land quality was observed, leading to increased crop yields.
- *Oorani:* Multiple-use tanks and ponds in small communities, for drinking water, aquaculture, irrigation and groundwater recharge. Restoration of these tanks provides a preferred source of drinking water, saving 365 hours or 45 working days per household that were spent on fetching water. Health is improved, and children attend school more regularly.
- *Wastewater:* Use of wastewater in peri-urban agriculture in semiarid areas provides water and nutrients and can be very productive. Pollution of surface water and groundwater is a drawback. Using the wastewater for nonedible cash crops can reduce some health risks.
- *Drip irrigation:* Small-scale drip irrigation for vegetables. Popular with women and good for food security. Utilizing this technology resulted in water savings of 50% and yield increases of 30 to 50%. Higher incomes have resulted in better nutrition.

Source: www.iwmi.cgiar.org/smallholdersolutions

4.5.2 Catchment management

Many small and medium-sized catchments in upper watershed regions encompass all the complexity of the smallholder landscape. The farming systems are diverse and include a mosaic of land uses, including production of annual crops, perennial crops, livestock, trees and non-timber forest products. Management of sloping and marginal lands in upper watersheds is problematic and is often unsustainable as currently practiced in many areas. In tropical parts of Asia, for instance, serious erosion caused by intense rainfall on sloping lands, which is exacerbated further by unsustainable land use patterns, is a major problem for agricultural development. Besides land loss and sedimentation, soil erosion also reduces crop yields. In Sub-Saharan Africa, for example, projected yield reduction due to erosion in 2020 is estimated at 14.5% (Lal 1995). IWMI research has highlighted the major crop production consequences of declining soil fertility due to soil mining and erosion (Drechsel *et al.* 2001) and evaluated methods to quantify consequences in economic terms (Enters 1998; Drechsel and Gyiele 2000; Drechsel *et al.* 2004).

Soil fertility is one of the key factors in determining agricultural output. In the African context, for example, soil nutrient depletion is the main biophysical factor limiting increases in per capita food production for the majority of small farms. Soil nutrients are also a primary, though often ignored, factor determining the costs and benefits of agricultural water management interventions and water productivity. Economic valuation of soil nutrients and their economic and ecological roles can be performed with methods based on replacement cost, productivity change, willingness-to-pay, hedonic pricing, and total factor productivity (see Drechsel *et al.* 2004). One particularly difficult area for economic valuation is soil organic matter (SOM), which is critical for soil productivity and water balance. Determining values to apply to SOM and soil carbon services is not straightforward as the following examples show.

In northern Europe, gardeners have used nutrient-poor and acidic peat over many years for soil structure amelioration and water retention. In Germany and Switzerland, they paid about $240–330 per metric ton of carbon. A shadow price for soil carbon of about $10 to 20 (ranging from $5 to 40) per metric ton of carbon emitted is thought to reflect a broad range of potential environmental damages caused by the loss of SOM. Carbon sequestration through agroforestry in African smallholdings suggests an estimated input cost (rock phosphate, tree seedlings and labor) of $87 per ton of carbon. Significantly lower costs (<$10) are possible via tropical tree plantations. Actual examples for credits for emissions abatement through carbon sequestration are in the range of $2 to 5 per ton (see Drechsel *et al.* 2004).

Although improving smallholder productivity at the farm level is one goal of catchment research, the longer-term focus is primarily on scales larger than the individual farmer, especially with an emphasis on the integration of the social and biophysical processes within complex landscapes (Greenland *et al.* 1994). From this perspective, IWMI research in the theme aims to understand aggregate effects on the landscape, including the on-site, off-site and downstream effects of soil erosion and land degradation. Obviously, capacity building to increase the ability of local partners for research at this scale and to promote an effective integration across disciplines remains a strong component of IWMI activities in catchment research.

IWMI catchment research comprises two large-scale and long-term projects, which were initially managed by IBSRAM, but incorporated later into IWMI programs in 2001. The projects are the *ASIALAND* Management of Sloping Lands (ASL) network and the Management of Soil Erosion Consortium (MSEC). In view of the long-term nature of these programs, it is relevant to review their scientific contributions to erosion process research, especially to see how these projects contributed to the evolution in soil erosion research that occurred in the late 1980s through the 1990s. In view of their network-based design, the data-collection and capacity-building impacts of these projects have been distributed widely throughout the region. The significant areas of research promoted by these projects include the linkages among erosion processes, farming systems and land use at the catchment scale, off-site impacts of erosion and their connection to public policy.

4.5.2.1 Erosion research: Evolving from technology to extension

IWMI and the former IBSRAM have a history of sloping land research that began in 1988, especially with the inception of the ASL network. Under this project, research was initiated in China, Lao PDR, Indonesia, Malaysia, Philippines, Thailand and Vietnam to quantify erosion under various land use practices, climates and soil types. This project evolved over time, in accordance with global thinking on how best to address problems of erosion on sloping uplands. The Swiss Agency for Development and Corporation (SDC) was its major donor and it took a keen interest in the conceptual developments in the network; partner institutions contributed substantially to the string of ASL projects. During 1988-94, the focus was on validation of technology for sloping lands with field experiments of practices for controlling soil erosion, including technology demonstration and training as well as research (IBSRAM 1992, 1995).

During 1995-97, however, there was a transition from research to extension and technology promotion with a larger focus on on-farm research to test technologies for economic potential and resource sustainability (IBSRAM 1998). Since 1998, the scope of this transition has been further broadened to emphasize the development and validation of methods for technology promotion, and the importance of enabling institutions. Throughout the project, on-site erosion mitigation was the primary focus. Contributions from this project included large multi-country data sets on erosion abatement potential, different soil-conservation techniques in the tropics, documentation of soil conservation benefits of agroforestry (contour hedgerows) from the ASIALAND and PACIFICLAND networks (Craswell *et al.* 1997) and extensive documentation of results by all network partners (IBSRAM 1998). Long-term experiments (Eusof *et al.* 2002; Sentaheunghoung 2002) demonstrated that soil organic matter and nutrients could be stabilized after 5 to 10 years of erosion mitigation treatments.

By 1998, these activities led to the establishment of the Management of Soil Erosion Consortium (MSEC), one of the four consortia created as part of the CGIAR system-wide initiative for Soil, Water and Nutrient Management (SWNM) (Craswell and Niamskul 2000). The project design captured newer thinking on INRM, in that research frameworks should ensure the participation of a range of stakeholders (Greenland *et al.* 1994) and include linkages between productivity-enhancing and resource-conserving research; and between research and the diffusion/adoption of research results at different spatial and temporal scales (Javier and Voss 2003).

Consultation meetings and dialogues were undertaken, and a network of research catchments, equipped for hydrology and soil erosion research, was established in six Southeast Asian countries (Indonesia, Lao PDR, Nepal, Philippines, Thailand and Vietnam). The catchments selected were 50-280 ha in size and receive between 700 and 2,200 mm of annual precipitation (Table 4.2). All catchments comprise different types of land use (forestry, tree crops, annual crops, bare land and grassland) and there are diverse stakeholders. A network of soil research institutes of participating countries carries out the field research.

Table 4.2 Characterization of the catchments in the MSEC research network.

Country	Indonesia	Lao PDR	Nepal	Philippines	Thailand	Vietnam
Catchment name	Babon	Huay Pano	Masrang Khola	Mapawa	Huay Yai	Dong Cao
Province	Semarang	Luang Prabang	Chitwan district	Bukidnon	Phrae	Hoa Binh
Elevation (m)	390-510	400-700	650-1400	1080-1505	400-480	125-700
Catchment size (ha)	285	67	124	84	93	45
Slope (%)	15-75	30-80	40-100	15-60	12-50	40-60
Rainfall (mm)	2500	1403	2200	2537	1077	1500
Vegetation and land use	Rice, maize, rambutan	Forest, bush fallow, rice, maize, job's tears	Forest, grasslands, rice, maize, millet, potato	Forest plantation, open grassland, maize, potato	Soybean, mung bean, tamarind	Cassava, rice, maize, taro, peanut
Households	405	80	54	70	50	38
Population	1812	427	354	155	3655	196
Land tenure	Owners, shareholders	State-owned, Land use right	Certificate of ownership leased	Private owner	Land use title	Land use right

Source: Maglinao and Valentin, 2003.

4.5.2.2 Erosion processes, farming systems and land use

The MSEC results helped to demonstrate the value of embedding process-oriented science into action research aimed at capacity building and impact. Detailed erosion-process research revealed that tillage erosion, a result of land preparation and repeated weeding operations increased with increasing slope in several different farming systems (Figure 4.1) and, was of the same order of magnitude as water erosion, significantly accelerating fertility transfer down slopes (Dupin *et al.* 2002).

Research within MSEC on weed ecology and invasive weeds in slash-and-burn agriculture (de Rouw 2001) demonstrated increasing weed pressure when natural fallows were shortened. These two factors result in a negative synergy whereby increased numbers of weeding operations exacerbated the problem of tillage erosion. Solutions lie in farming systems such as improved fallows, or no-till mulch systems that reduce weed pressure, reducing tillage operations and tillage erosion. Thus, production systems that reduce farm labor requirements coincide with reducing erosion on these sloping lands (Table 4.3 and Figure 4.2). Studies of nutrient losses from MSEC catchments and the on-site costs of soil erosion not only produced quantitative findings but also served to increase the appreciation of the value of soil conservation (Chaplot *et al.* 2002; Janeau *et al.* 2003b; Chaplot *et al.* 2004).

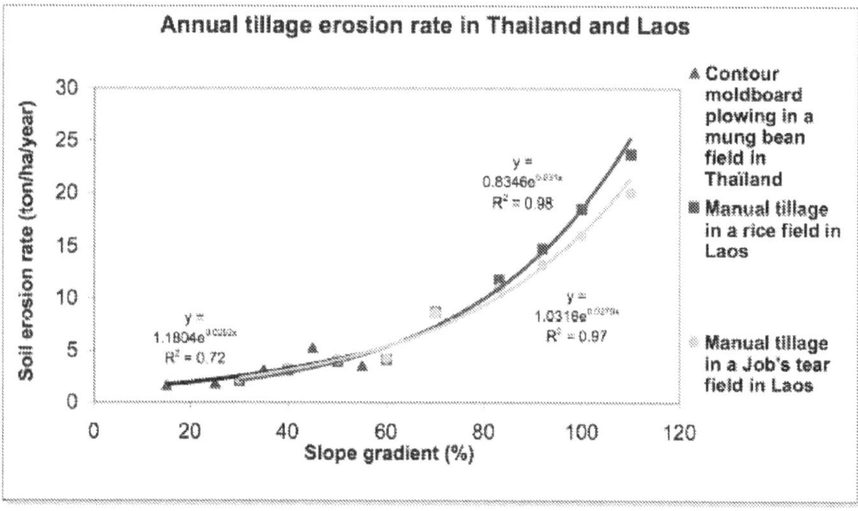

Figure 4.1 Relation between erosion rate and slope across farming systems in Thailand and Lao PDR.

Table 4.3 Impact of farming system on sediment yield and labor use in upland rice cultivation, Houay Pano catchment, Lao PDR.

Farming systems	Catchment area (ha)	Bedload (tons/ ha)	Suspended sediment load (tons/ ha)	Total sediment yield (tons/ ha)	Labor requirement (days/ ha)
Slash-and-burn (control)	0.62	4.74	0.99	5.74	210
Improved fallow	0.64	0.4	0.01	0.42	171
Improved fallow + Contour planting	0.57	1.95	0.56	2.51	182
Mulch + No tillage	0.73	0.11	0.47	0.58	138

Source: Maglinao and Valentin 2003.

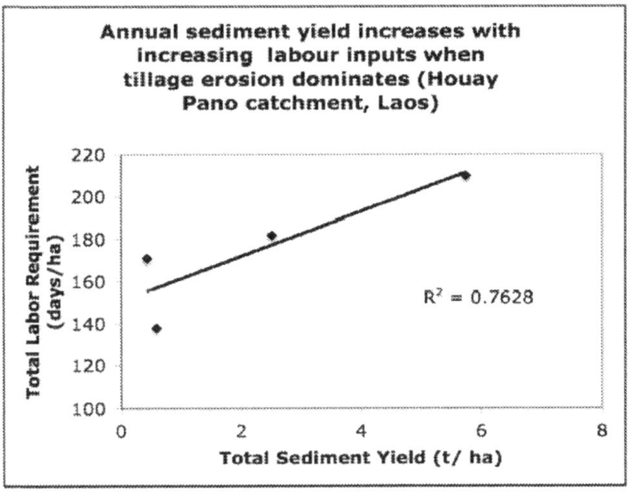

Source: Maglinao and Valentin 2003.

Figure 4.2 Relationship between sediment yield and labor needs across farming systems in the Houay Pano catchment, Lao PDR.

Catchment studies have also provided significant results with respect to larger-scale processes and links between land use and erosion. The research gave more insight into the large temporal and spatial variability of erosion, and its relation to soil surface, slope (Janeau *et al.* 2003a), soil cover and rainfall pattern. Surprisingly, it was found that there can be less runoff than predicted on steeper slopes when slopes are convex, because the topography results in natural infiltration traps (Janeau *et al.* 2003a). Also, most of the eroded soil does not move out of the catchment where it originates, but is deposited a little further downstream, which makes 'scale' an integral component of measures for runoff and sedimentation. Catchment modeling efforts took several forms. A detailed model was initially used to study sheet and rill erosion in plots, with ACIAR support.

Calibrations with field observations revealed how much the erosion potential of the soil surface changed during the season (Yu and Rose 1999; Yu *et al.* 1999). A deterministic model (MSEC1) was developed and parameterized for the Lao PDR catchment (Chaplot *et al.* 2002). It showed that in this catchment, on-site soil erosion increased fourfold, when the fraction of the area under annual cultivation increased from 9 to 60% and it suggested that the fraction of land cultivated was a good indicator of catchment sediment yield. The model was used to demonstrate that land use affects soil erosion more than climatic variability or predicted climatic change (Chaplot *et al.* 2002). For this explanatory model, it is necessary to have rainfall data with intervals as short as 6 minutes. A simpler model was derived from MSEC1, using decadal precipitation values as input, for application at a wider scale.

4.5.2.3 Off-site impacts of erosion

Off-site or downstream impacts of erosion, including the siltation of irrigation canals, reservoirs and tanks, the degradation of aquatic ecosystems and the pollution of drinking water are potentially greater than on-site impacts. One way to evaluate these off-site effects is through the estimates of dredging obtained from irrigation canals. For instance, in the Manupali River Irrigation System in the Philippines, it is estimated that a total of 84,685 m^3 of sediments have been transported into the system since 1995 (Ilao *et al.* 2002). The effects of such siltation are also serious. For instance, the evaluation of sedimentation in the case of the Mae Thang Reservoir in Thailand indicated a 10% reduction in the storage volume just within 7 years of its operation (Janeau *et al.* 2003b). As a result, instead of a 100-year life span initially estimated for this reservoir by the Royal Irrigation Department, current estimates indicate a much shorter life span of only 70 years (Inthansothi *et al.* 2000).

The impacts of sedimentation are still more serious at an aggregate perspective at the global level. It is predicted that more than 25% of the world's water storage capacity will be lost in the next 25 to 50 years in view of the absence of measures to control sedimentation in both large and small reservoirs (Palmieri *et al.* 2001). In addition to effects on the life of the reservoir and its irrigation service area, erosion may also degrade aquatic ecosystems. During the first heavy rains, sediments laden with pesticides used in upland agricultural systems may affect fish catches (Boonsaner *et al.* 2002).

4.5.3 Rehabilitation of degraded lands

Since the degradation of land and water resources in terms of salinization, erosion, nutrient mining and chemical pollution reduces the productivity of both land and water resources, it undermines the global capacity to produce food, livelihoods and environmental services. In the face of these challenges, many smallholder farmers find it more and more difficult to earn an income and achieve food security. They have no choice but to continue to live with the consequences of further resource degradation unless research and policy efforts are redoubled to arrest their social, economic and environmental consequences. It is the urgency of these problems combined with the tremendous potential that exists for improving the productivity of rain-fed systems that motivates and guides IWMI research under this subtheme.

On the research front, many reports have been published in recent years on the trends of global water and land resources and their impacts on agriculture, food security, poverty and environment (e.g., Scherr and Yadav 1996; Wood *et al.* 2000; Bridges *et al.* 2001). In particular, IWMI has developed a set of recommendations for policy and research, based on its own work as well as from a synthesis of other major works (Penning de Vries *et al.* 2002b). IWMI research in this area falls into three main categories: (a) assessment work focused on learning from positive examples of reversing degradation; (b) farm-scale rehabilitation of light-textured tropical soils; and (c) understanding regional and global processes that contribute to land degradation.

4.5.3.1 Learning from 'bright spots'

While the overall general picture of resources degradation is bleak, suggesting downward spiraling trends, there are also many 'bright spots', where small or larger communities have taken a different development path and raised their incomes and food security and improved the quality of their natural resources. These 'bright spots' are a significant area for research aimed at both documenting practical experiences of resource conservation and management as well as developing key lessons for reversing the downward trends in resource degradation (Penning de Vries *et al.* 2002b). They are also the sites with sustainable farming (Pretty and Hine 2004). In recent years, there have been many organizations, which focus on 'success stories' as practical sources for policy guidance.[7]

[7] Groups with projects cataloguing and detailing success stories: Centre for Development and Environment (CDE), Berne; Centre for Environment and Society, University of Essex; Ecoagriculture Partners; FAO Land and Water Development Division; FAO/AGL Gateway Project; Ingenious Farmers; Centre for International Cooperation, University of Amsterdam; IRCD; Sustainability Institute, Stockholm Environment Institute (SEI);

IWMI has adopted the innovative framework of 'bright spots' for systematically looking into the factors that initiate and sustain the best practices and evaluating their impacts on land and water resources both on and off the sites. The thrust of IWMI research in this respect is how such local successes can be repeated elsewhere and scaled up to basin-wide contexts. A summary of 'bright spot' cases compiled by the IWMI project makes it clear that yield improvements are possible across a wide range of farming systems (Table 4.4). It is also clear that major increases were found where yields were low (Figure 4.3), i.e., areas where the yield gap between potential and actual production is greatest, lending strong support for the optimistic view that yield gaps can represent areas of opportunity, rather than the negative approach of dismissing them as hopeless cases.

Table 4.4 Summary of adoption and impact of sustainable technologies and practices in 286 projects in 57 countries.

FAO farming system category	Number of farmers adopting	Hectares under sustainable agriculture	Average % increase in crop yields
1. Smallholder irrigated	179,287	365,740	184.6 (±45.7)
2. Wetland rice	8,711,236	7,007,564	22.3 (±2.8)
3. Smallholder rain-fed humid	1,704,958	1,081,071	102.2 (±9.0)
4. Smallholder rain-fed highland	401,699	725,535	107.3 (±14.7)
5. Smallholder rain-fed dry/cold	604,804	737,896	99.2 (±12.5)
6. Dualistic mixed*	537,311	26,846,750	76.5 (±12.6)
7. Coastal artisanal	220,000	160,000	62.0 (±20.0)
8. Urban-based and kitchen garden	207,479	36,147	146.0 (±32.9)
All projects	12,566,774	36,960,703	83.4 (±5.4)

Note: Yield data are from 405 crop-project combinations; Standard errors are in brackets. * indicates mixed large commercial and smallholder farming, mainly in southern Latin America.
Source: Pretty *et al.* 2004.

To derive a generic understanding of the success of these individual bright spots, a preliminary drivers' analysis was undertaken in which the relative importance of a range of individual, social, technical and external drivers (Box 4.2) was determined (Noble *et al.* 2004). Case studies were classified into three primary groups: Community bright spots such as integrated watershed development, in which investment in social capital such as community

---

UNEP success stories; WOCAT, World Overview of Conservation Approaches and Technologies, Berne (not a comprehensive list).

organizations was as important as technical inputs for success and sustainability; Technology bright spots, which were successful largely through strong individual initiative and because the new technology or knowledge was particularly appropriate and effective; and Spontaneous bright spots, where significant improvement was made in resource conditions and profitability without external investment, driven by strong leadership and the availability of appropriate technology. It is hoped that this type of analysis of drivers will help inform efforts aimed at replicating success.

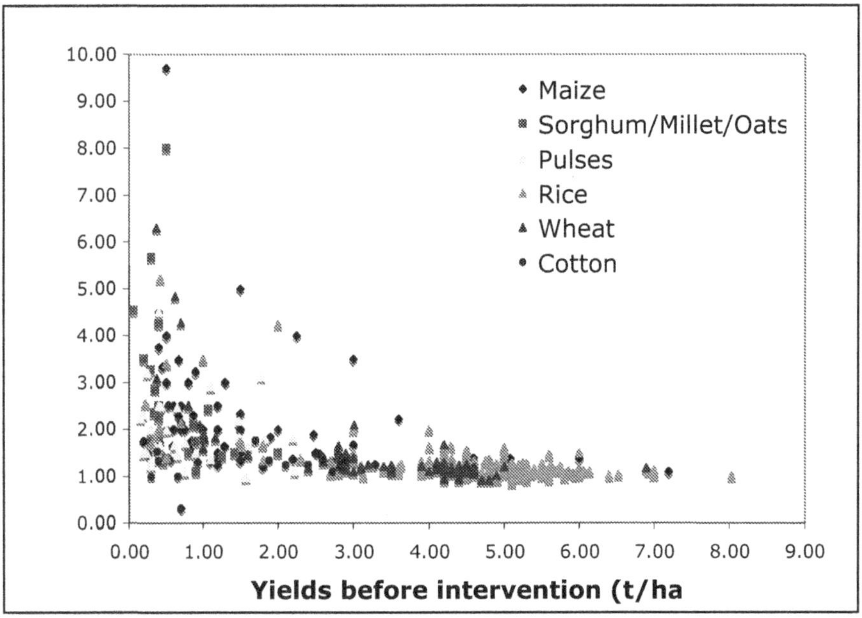

Note: Results cover 360 crop yield changes in 198 projects.

Source: Pretty et al. 2006.

Figure 4.3 Changes in crop yields with sustainable technologies and practices.

In arid and semiarid regions, salinization of soil and water is a key degradation issue. It affects productive potentials, reduces water use efficiency, results in loss of high-quality water to saline sinks and generates abandonment of previously arable lands. Although data are poor, estimates indicate that, worldwide, 20% of irrigated land suffers from salinization and waterlogging (Wood *et al.* 2000). The bright spots approach was applied in Central Asia to document the characteristics of farming enterprises that are more successful than their neighbors at maintaining soil quality and farm productivity in the larger

Box 4.2 Key drivers for success of Bright Spots.

1. *Aspiration for change*. This reflects an internal demand by an individual or community for change that may be driven by faith or a wish to try something different.
2. *Leadership*. In order for a bright spot to develop and continue there is a need for strong leadership. This may include a single individual or a group that champions change. Social
3. *Social capital*. Bright spots develop where there are community organizations, networks and partnerships (private as well as public). This social capital also includes intangible aspects of social organizations such as norms and rules of behavior that can play an important role in promoting sustainable change.
4. *Participatory approach*. Bright spots require deliberative processes that actively involve the community in the decision-making process. This includes a strong element of learning and teaching. Technical
5. *Quick and tangible benefits*. Immediate tangible benefits to the community or individual are an important requirement for the development of a bright spot. For example, this may include increased yields within the first year of implementing changes, a reduction in the costs of labor, etc.
6. *Low risk of failure*. Resource-poor farmers by their very nature are risk-averse; hence any changes made to create a bright spot need to have an element of low risk.
7. *Innovation and appropriate technologies*. Innovations, new technologies and information are important key components in the development and continuance of a bright spot. This includes new skills and knowledge that contributed to the development of a bright spot.
8. *Market opportunities*. In order for a bright spot to develop, markets need to be present, accessible and assured to effect change.
9. *Property rights*. For the development and continuance of a bright spot, secure (individual or communal) land user rights and tenure are important to facilitate change.
10. *Supportive policies*. Favorable changes in supportive policies at the local, regional and national levels are key drivers for the development and continuance of bright spots.

context of land and water salinization (Hassan *et al.* 2004). In this case, skill and leadership appeared to be key drivers of change. These cases clearly indicate that on-farm improvements in management can have a significant positive impact on profitability and sustainability of farming systems that are affected by salinity. In terms of technology innovation, mechanical approaches are promising. For example, using a simulation model in Pakistan, it was demonstrated that if the water table is below 1 m depth, pre-monsoonal cultivation (that enhances infiltration of rainwater) for a few years can reclaim salinity-affected areas; the new situation is sustainable if the treatment is repeated every year (Prathapar and Qureshi 1999).

4.5.3.2 On-farm rehabilitation of degraded soils

A common view is that smallholder farmers can only improve the productivity of their farm slowly, particularly if the area is already degraded. However, strong examples to the contrary exist, such as cases from Africa where smallholder farmers were very successful using an integrated variety of land and water conservation approaches to significantly improve the condition of their land and water resources (e.g., Auerbach 1999). IWMI's work in Northeast Thailand is another example of rapid reversal of degradation (Noble *et al.* 2004). This work is exciting in that it shows applying particular clays to chemically degraded soils can rejuvenate them fully. These light-textured soils had been overexploited for many years, such that 80% of the original nutrients had been lost. Soil carbon was also lost resulting in decreased water-holding and nutrient-holding capacity.

Clay additions increased the water and nutrient use efficiencies of crops on these soils, increasing productivity more than threefold (Figure 4.4) in the first year after application (Noble *et al.* 2004), while the soil water-holding capacity increased from 14 to 18% (Suzuki *et al.* in prep) (Figure 4.5). Demonstration of the results of using clays at farmers' field days evinced much enthusiasm from farmers, researchers, NGOs and regional planners, and field evaluations are now being carried out with, and by, them. Where feasible, clay rejuvenation may become an important component of rehabilitation for light-textured tropical soils that are found on more than a billion hectares worldwide. This option will be further researched and the synergistic effect of combining the application of such clays with precision irrigation methods explored.

4.5.3.3 Regional and global degradation processes

It is also important to look at nutrient recycling, and lack thereof, at larger scales in relation to soil degradation. It becomes quickly apparent that lack of recycling is more common than recycling. Within farms, fertility often becomes concentrated on the best fields (a relocation of resources that increases farm efficiency). Since antiquity, village livestock have provided nutrients in the form of manure for common grazing lands. But significant amounts of nutrients are also exported from farms and communities in the form of harvested products.

Recent research has shown that this leads to significant losses of soil nutrients: As much as 10% of the value of the total agricultural product would be needed to replace the lost nutrients (Smaling 1993; Drechsel *et al.* 2001).

There is scant recognition that unbalanced nutrient fluxes between urban centers and rural areas pose a serious threat to the future.

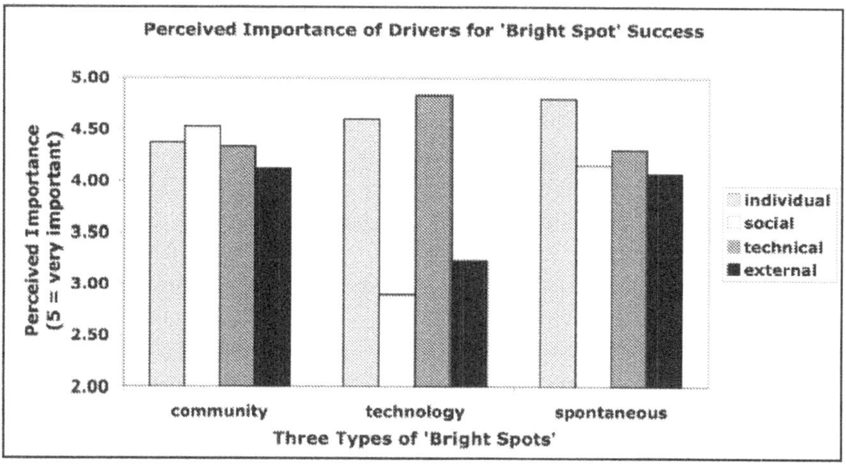

Note: This analysis is based on 15 community-based, 95 technology-based, and 3 spontaneous bright spots.

Figure 4.4 Preliminary drivers' analysis for three types of bright spots.

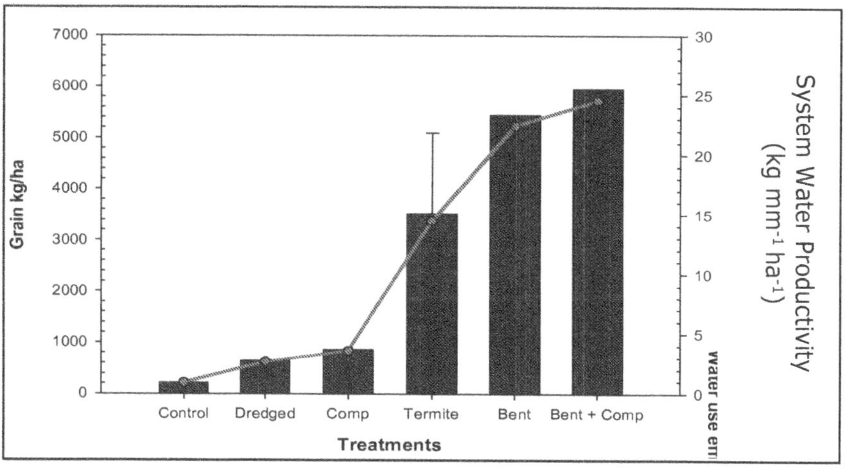

Figure 4.5 Yield and water productivity response of forage sorghum to soil treatments in degraded soils in Northeast Thailand.

Large cities, particularly megacities, consume large quantities of food and hence large quantities of nutrients. Of these nutrients, very little is recycled to agriculture through wastewater or compost; most flow into the sea or ocean or pollute soils in and around the city. We calculated how many nutrients are annually imported in a large city like Bangkok and what happened to them. Recycling is minimal, waste treatment quite limited and most nutrients either flow as wastewater into the sea or are withheld in the city environment (Faerge *et al.* 2001). Strong concentrations of nutrients in cities and then in rivers on the one hand, and depletion in marginal agricultural areas on the other are poles of an untenable ecological imbalance. With the expansion of megacities and the rise of income of their populations, this imbalance is likely to grow and will lead to major health and environmental problems. Within IWMI, this line of research has been pursued under peri-urban and urban agriculture and use of wastewater (Drechsel *et al.* 2001) and analysis of urban-rural nutrient flows (Penning de Vries forthcoming).

4.6. CONCLUSIONS

IWMI's Theme, Smallholder Land and Water Management, was from the beginning founded on the need for synergistic research, integrating land and water expertise. For a long time, water research and soil research have been dealt with in separate research institutes, professional societies and government departments. Merging with the programs of IBSRAM made this ideal of synergy a concrete reality for IWMI, such that IWMI is now recognized for its integrated land and water research.

In addition to the myriad publications and impacts already cited and discussed under the subthemes, some broad contributions were made to science and capacity building. We produced knowledge for a broad readership and a range of books and workshop proceedings. Some examples of these publications include Saktivadivel and Sally 2003, who characterized different types of water and irrigation management in relation to climates; included in the *Encyclopedia of Water Science*, this publication will reach many professionals. Renault and Godaliyadda (1999) generalized typology for gravity irrigation systems that can help to categorize and compare management approaches. Keller *et al.* (2000) reflected on benefits and costs of scales of water storage facilities ranging from large dams to medium-sized tanks.

The Theme objectives were to find and promote ways in which smallholders can improve their productivity and income through improved management of water and land resources. Innovations studied include methods for land and water conservation on sloping lands and in catchments (Southeast and South Asia); treadle pumps (Asia, South Africa); methods for improving access to water (India); methods to rejuvenate soils (Southeast Asia); rainwater harvesting on homesteads (South Africa); and the emergence of bright spots. We learned that the opportunities for rehabilitation of water and land resources by smallholders are exciting and offer much scope for further development.

Finally, IWMI's Theme, Smallholder Land and Water Management, highlighted several issues for the international community. Perhaps most significantly, IWMI's research has highlighted the important contribution of rain-fed agriculture to food security and the real and achievable opportunities that exist for increasing land and water productivity in rain-fed areas. The lessons from these and other studies have helped to shape the Institute's future research on smallholder land and water management under the new theme: Land, Water and Livelihoods.

5

Sustainable Groundwater Management

Tushaar Shah

5.1 INTRODUCTION

Where agricultural growth has been the driver of rural poverty reduction and improved food security in recent decades, particularly in Asia, intensification of groundwater irrigation has emerged as a major socio-technical reality. However, the importance attached to the study of social dimensions of this reality is far from commensurate with its growing significance. True, after all, as part of the hydrological cycle, groundwater is no different from surface water. However, the behavior of groundwater - and groundwater institutions - is different from that of surface flows and canal irrigation institutions in complex and material ways; and yet, almost all irrigation management literature during the 1980s focused on large public irrigation systems based on surface water. For instance, less than 5% of the research produced by IIMI during its first decade dealt with groundwater management even though the share of groundwater in irrigated areas in some of its priority countries - Pakistan, India, Bangladesh, Nepal, China - had far surpassed the share of surface systems. Further, the British Geological Survey, a leading groundwater research organization in the world

© 2006 IWMI. *More Crop per Drop*. Edited by M.A. Giordano, F.R. Rijsberman, R. Maria Saleth.
ISBN: 9781843391128. Published by IWA Publishing, London, UK.

got its first PhD in social science only in the new millennium; the US Geological Survey does not have one even now.

Intensive groundwater irrigation has generated totally different socioeconomic and institutional dynamics compared to surface irrigation. Surface irrigation almost everywhere is in the public sector; groundwater irrigation is invariably in the private sector. The former is therefore 'formal' in the sense that public agencies are, at least in theory, in charge of 'managing' it;[1] the latter has created a totally informal economy, largely untrammeled by regulatory frameworks. The former is often a result of planned development with the help of government and donor investments; the latter is largely unplanned, and energized by farmer investments that are only indirectly influenced by public policies such as for institutional credit and targeted subsidies. The agencies dealing also with groundwater are almost everywhere, differing from those managing canal irrigation, with little or no coordination between them.

The key management challenges facing surface irrigation - such as improving water productivity, O&M cost recovery, making participatory irrigation management work, dealing with waterlogging and salinization and struggling with head-tail inequity (Merrey 1997) - are qualitatively different from the management challenges facing groundwater - such as regulating runaway growth in groundwater structures, arresting groundwater depletion, coping with contamination and protecting aquifers, ensuring equitable access to a common property resource, and managing energy use in groundwater pumping (IWMI 2003a). In the management between surface systems and groundwater systems, then, only water is the same; all else are different. As a result of these differences, a student of water management who focuses on surface irrigation is sure to miss out on more than half of the action on the ground.

Understanding sustainable groundwater management in the developing world requires blending three distinct perspectives: (a) the resource perspective; (b) the user perspective; and (c) the institutional perspective. IWMI's Sustainable Groundwater Management (SGM) theme rests on the premise that global knowledge development as well as capacity building on groundwater use in agriculture so far is dominated by the 'resource perspective'; and a critical value-adding contribution is to be made in expanding global knowledge and capacity in user and institutional perspectives. The larger goal that has driven SGM research is to contribute to the sustainable use and management of

---

[1] A set of recent studies on major, medium and minor surface irrigation systems in different parts of India suggests that beneath this veneer of 'management', its operation is anarchic (Neetha 2004; Vashishtha *et al.* 2003; Rajagopal 2003; Joy and Paranjape 2003; Meher 2003; Patil and Doraiswamy 2003; Shah, Anil 2003).

groundwater and ensure, thereby, the food and livelihood security of the poor in Asia and Africa. In this direction, effort is applied to three key purposes, each a subtheme with a coherent cluster of activities woven around it: (a) to develop and disseminate a more accurate and refined understanding of the socio-ecological value of groundwater, and the nature and scale of the consequences of its unsustainable use; (b) to identify and promote research on promising technologies and management approaches with potential to help achieve sustainable groundwater use; and (c) to identify, through applied policy research, a toolkit of 'sustainability solutions' and aggressively promote these amongst strategic players in national and regional groundwater systems. The remainder of this chapter tries to assess how far SGM research in IWMI has been able to achieve these purposes.

5.2 GROUNDWATER SOCIO-ECOLOGY OF SOUTH ASIA

Within this first subtheme, the SGM theme has tried to further develop and refine the 'big picture' of the groundwater economy. The work has primarily focused on South Asia and North China by examining the changing landscape of groundwater irrigation in the two subregions and from that designing a four-stage model to characterize the process of groundwater intensification. Based on the findings from these two subregions, SGM researchers have also examined the implications for Asia as a whole as well as for Africa. It is important to note that we present here IWMI's evolving understanding of the big picture of global groundwater use in agriculture. However, even as this book goes to the press, we have new studies that significantly modify our understanding of the scale, reach, structure and functioning of the groundwater economy of North China Plains. In a field of knowledge that is evolving rapidly, assessments of the nature presented here need to be taken with the degree of tentativeness they invariably contain.

5.2.1 Getting the big picture right

A key task of creating a 'big picture' of the groundwater economy began with IWMI's development of a discussion paper on 'Global Groundwater Situation' for the 2nd World Water Forum (Shah, Molden *et al.* 2001). Subsequently, a new body of IWMI literature on groundwater socio-ecology has helped add more substance and nuance to the emerging 'big picture'. Some of this work was summarized in a paper for the 3rd World Water Forum (IWMI 2003b). This work has so far concentrated on using existing and new databases to assess the quantitative trends in groundwater use in agriculture; but a 2002 IWMI survey of nearly 2,600 tube well owners in 290 villages covering all of India, Pakistan except Balochistan, 20 districts of Nepal terai and Bangladesh has helped deepen this understanding. A similar survey of North China Plains is under way.

In South Asia and North China, where governments and international donors have invested heavily in public-sector irrigation projects during the 1960s and 70s, groundwater irrigation has become the mainstay of agriculture. Farmer-financed groundwater irrigation capacity has caught up-and in some regions, even surpassed-public irrigation. In India, out of some 20 million farmers[2] one in every five already owns a tube well; and during the coming years, this number will likely grow at a rate of around 0.8-1 million/year. Moreover, through localized, fragmented pump rental markets, a representative pump owner sells irrigation service to an average of three neighboring plots (Mukherji and Shah 2002), implying that finding Indian farmers who do not use groundwater irrigation *at all* would not be very easy at the turn of the millennium. This is also the case with Pakistan Punjab and North China (Shah *et al.* 2003; Mainuddin 2002). In the first such attempt to estimate the size of the groundwater economy, it was calculated (Table 5.1) that the market value of groundwater used in these regions is likely to be around $15 billion/year;[3] and we estimate that the value of agricultural output made possible by it is likely in the range of $40-60 billion/year. By global standards, this may not amount to much. However, in poor agrarian economies, this makes groundwater irrigation a big business, and an even bigger livelihood generator (Mainuddin 2002).

5.2.2 South Asia: Public investments vs.private initiative

The early 1970s seem to have been the watershed in groundwater development (Figure 5.1), before which, agricultural expansion in India was driven by public investments in large surface irrigation projects. But, 20 years later, we have witnessed a massive shift. An Irrigation-Agriculture Macroeconomic model explaining factors that contributed to agricultural productivity variations across 252 districts in India, representing 85% of the country showed that during 1971-73, surface water irrigation (SWI) explained the bulk of the variation. During

---

[2] These are estimates made by the SGM theme but will soon be verified by a new Minor Irrigation Census completed by the Government of India for 1999-2000.

[3] Pumping costs understate the value farmers place on groundwater because throughout South Asia, farmers without own wells routinely buy water from well owners at prices that reflect substantial premia over pumping costs. These prices, gleaned from socioeconomic surveys, are taken to be more reflective of farmers' valuation of groundwater. However, these prices are nothing but rental value of wells and pumps which are a great deal cheaper in China than in South Asia. As a result, the valuation of groundwater use in China in this method is lower than in South Asia but its productive contribution is much larger.

1991-93, it is groundwater irrigation (GWI) that largely accounted for the variations in the value of agricultural output/ha (Deb Roy and Shah 2002).

Table 5.1 The size of Asia's groundwater economy.

No	Particulars	India	Pakistan Punjab	Bangladesh	Nepal Terai	North China Plains
1	Wells (million)	20	0.5	0.8	0.06	4.5
2	Average output/well (m^3/hr)	25	100	30	30	30
3	Average hours of operation/well/yr.	330	1090	1300	205	800
4	Price of pump irrigation ($/hr)	1	2	1.5	1.5	0.96
5	Groundwater used (km^3)	215	54.5	31.2	0.37	106
6	Value of groundwater used/year in $ billion	8.6	1.1	1.6	0.02	2.5

IWMI's research in this area has helped refine the big picture by substantiating existing hypotheses with new data and teasing new relationships and implications. In the North Indian context, for example, Dhawan (1982) suggested that the Green Revolution followed the tube well revolution in Punjab, Haryana and western UP. With new analyses in eastern India (Shah 2003b; Ballabh and Chowdhury 2003; Pant 2004) and at an all-India level (Deb Roy and Shah 2003), IWMI's research suggested the broader validity of the role played by tube well irrigation in catalyzing the Green Revolution throughout the subcontinent. Likewise, IWMI's work highlighted the changing relative contribution of land and water in agricultural value added. Fifty years ago, when ownership of farmland was the predominant source of rural wealth, land reforms were viewed as the chief redistributive instrument in the hands of the government. Today, thanks to pervasive spread of groundwater irrigation followed closely by Green Revolution technologies, improving access of the poor to groundwater has emerged as a potent instrument of redistributive policies (Chambers *et al.* 1987; Shah 1993; Deb Roy and Shah 2003).

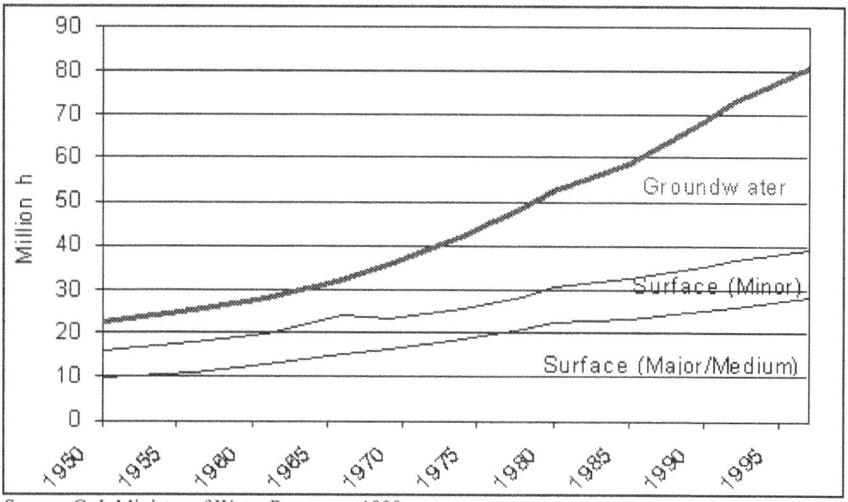

Source: GoI, Ministry of Water Resources 1999.

Figure 5.1 Iirrigated area by source in India.

Groundwater irrigation has thus given a human face to the Green Revolution, which would have remained confined to less than 20% of India's farmlands, had public canals been the only source of irrigation. Researchers noted long ago that a cubic meter of groundwater contributes more to yield/ha compared to a cubic meter of flow irrigation (Dhawan 1985). However, a major reason behind aggressive political support to groundwater development is that it is proving to be a powerful 'equalizing' influence. While access to canal irrigation is limited to those located in command areas, there is more equality in access to groundwater irrigation compared to the distribution of farmlands. Analyses of the Indian data gathered by a national census of 1993-94, shows that the distribution of ownership of diesel pumps and electric pumps is less unequal compared to the distribution of ownership of farmland (Deb Roy and Shah 2002). Lorenz curves for proportion of area irrigated by groundwater and surface water for 375 districts of India show that access to surface irrigation is more unequal compared to access to groundwater irrigation.

IWMI's work on the big picture has also drawn attention to the threat that much spread of groundwater irrigation in the subcontinent may be environmentally unsustainable because it is dependent only on natural recharge from rainfall and from limited surface runoff (Deb Roy and Shah 2002). Intensification of groundwater irrigation in areas with poor aquifers may help farmers irrigate crops initially; but soon, problems of groundwater depletion set in motion a costly chase of falling water tables, leaving a trail of dried-up wells and impoverished farmers. The process of 'creative destruction' through which groundwater intensification has occurred has seldom followed the tenets of hydro-geological wisdom. But then, without groundwater development, agriculture would have stagnated or declined in peninsular India and North China; food security would, of course, be endangered; but a more critical problem would be supporting rural livelihoods during several decades that these regions would take before a sufficient proportion of their agrarian populations could be transferred to off-farm livelihood systems (IWMI 2003b). The IWMI South Asia survey helped add depth and nuance to this big picture analysis of the groundwater socio-ecology of South Asia (Mukherji and Shah 2002; Shah *et al.* 2006). Some of its key findings are:

First, there has been a major change in the landscape of groundwater irrigation: Regional patterns of groundwater development are not in sync with the regional groundwater resource position. Rather, tube well density seems to closely follow population density and population pressure on agriculture. Thus, compared to large public irrigation projects, which are driven by hydrologic opportunity, groundwater development is proving to be more 'egalitarian'. Over 50% of groundwater structures surveyed were built only after 1993, suggesting that South Asia's 'groundwater juggernaut' is still accelerating (Figure 5.2). 56% of the irrigated land in survey villages are accounted for by 'pure groundwater irrigation'; counting conjunctive use, tube well irrigation's contribution rises to over 75% of net irrigated area while pure canal irrigation covers only 14.5% of irrigated areas. Most groundwater irrigation in South Asia is supplemental irrigation, with tube wells providing 25-35% of the total crop water requirements. Coarse cereals, pulses, oilseeds and fibers are preferred crops in pure groundwater irrigated areas; paddy-wheat cycle is the preferred pattern in canal and conjunctive use areas.

Second, the groundwater-irrigation nexus has become stronger. Despite huge and perverse power subsidies for groundwater irrigation, the diesel pump is the mainstay of the poor. Farmers respond strongly to energy pricing and supply policies, which vary substantially across the South Asian region; and as expected farmers with subsidized electricity use significantly more energy (and likely groundwater) per hectare compared to owners of diesel pumps or owners of electric pumps in regions without real electricity subsidies. Despite subsidized electricity and canal irrigation, it is the diesel pump that has by far the largest quantitative impact on explaining variations in Net Irrigated Area in 292 sample villages. While much as the world worries about groundwater depletion, salinity and reduced aquifer yields, for the South Asian farmers surveyed, the most critical problem facing groundwater irrigation is high energy cost of pumping followed by unreliable supply of electricity.

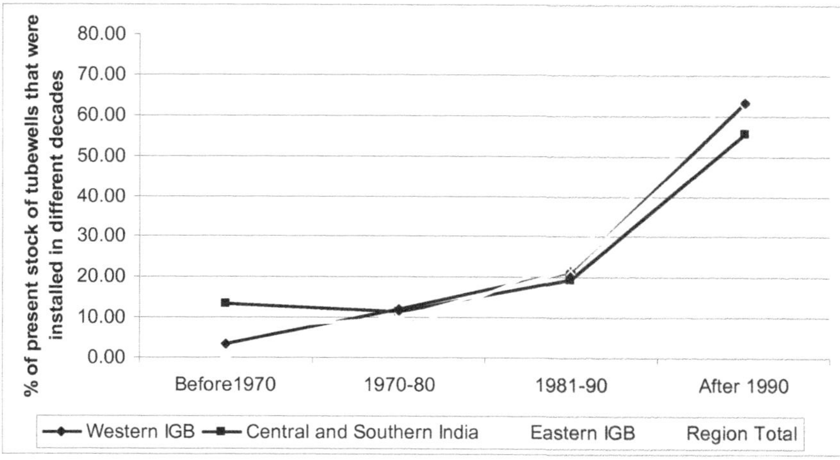

Figure 5.2 Pace of growth of tube wells in different parts of India.

Third, there have also been major changes in the nature and evolution of groundwater markets. Groundwater markets seem to have shrunk in Western and southern India (compared to the 1980s) but they have boomed in eastern India, Bangladesh and Nepal terai, besides Pakistan Punjab. Far from an expropriating 'water lord', the average South Asian water seller is a smaller farmer with more fragmented holding compared to non-sellers. Manufacturers of leading pump brands, who have penetrated the pump markets of South Asia, can be important informal sector partners to improve energy and water use efficiency in the South Asian groundwater economy.

All in all, agricultural use of groundwater in some of Asia's densely populated food baskets has experienced runaway growth after 1970. If current trends continue, the irrigation scenario of these regions will soon change beyond recognition. In India, for instance, at a compound annual growth rate of 3.81%/year, growth in groundwater irrigated area has contributed over 80% to the 2.2%/year growth in overall irrigated areas, with surface irrigated areas having stagnated since the mid-1980s. At these rates, by 2025, India will have nearly 100 m ha of irrigated areas but of these, over 82 m ha will be irrigated by groundwater wells, with surface irrigated area falling in absolute terms (Shah 2003a). Pakistan's Punjab and parts of North China Plains too have been experiencing similar exponential growth in private tube wells (Shah *et al.* 2003). While the rate of growth of groundwater draft is certainly lower than that of tube well numbers, none of these trends suggests that growth in groundwater irrigation is tapering off (Shah *et al.* 2003).

Despite all the productivity and livelihood benefits of groundwater irrigation, this runaway growth presents a frightening prospect because it will magnify manifold the negative externalities of groundwater overdevelopment viz., the rising cost of chasing a perennially declining water table, lost wetlands and biodiversity, and reduced low flows (Seckler *et al.* 1998b; Moench 1992; 1994b; Moench and Burke 2002). There are also reports of alarming rates of failure of wells due to falling water tables in western and peninsular India. Yet, there is no unanimity of views on sustainability of current patterns of groundwater use in South Asian agriculture. Some Western observers (e.g., Llamas *et al.* 1992; Llamas and Custodio 2003) argue that groundwater depletion is a minor problem affecting small pockets. It seems that this was perhaps true a decade ago but many of the small pockets are growing rapidly to engulf big regions.

5.2.3 Rise and fall of groundwater socio-ecologies

So regular and predictable is the process of groundwater intensification in a region and its decline due to overdevelopment of the resource that IWMI has proposed a four-stage model to characterize this process (Figure 5.3) and Table 5.2. This model underpins the typical progression of a socio-ecology from a stage where unutilized groundwater resource potential becomes the instrument of unleashing an agrarian boom to one that goes overboard in exploiting the resource. The highlight of the model is the lag with which public policy changes its posture; governments keep supporting groundwater intensification long after the need to regulate further development becomes imminent. This simple model has important lessons for developing regions of the world where groundwater intensification is yet to begin. The critical issue for Africa, for instance, is: can it put in place an institutional and policy framework that can generate a steady-state equilibrium, which sustains the groundwater-induced agrarian boom without degrading the resource itself? For Asia, the critical issue is: what might be done to sustain groundwater socio-ecologies under threat and keep them from falling over the edge of the precipice?

The answer, of course, depends on many factors including physical geography. For example, in humid alluvial plains of the Ganga-Brahmaputra-Meghna, Mekong and Yangtze Basins, vast reserves of unutilized groundwater resources offer a major opportunity for agricultural growth and poverty reduction. In contrast, the arid alluvial plains of the Indus and Yellow Rver Basins and the hard-rock regions of peninsular India pose significant management challenges. Table 5.3 outlines the groundwater governance challenge in different parts of Asia.

Stages	Stage 1	Stage 2	Stage 3	Stage 4
	The rise of Green Revolution and tube-well technologies	Groundwater-based agrarian boom	Early symptoms groundwater overdraft	Decline of the groundwater socio-ecology with immiserizing impacts.

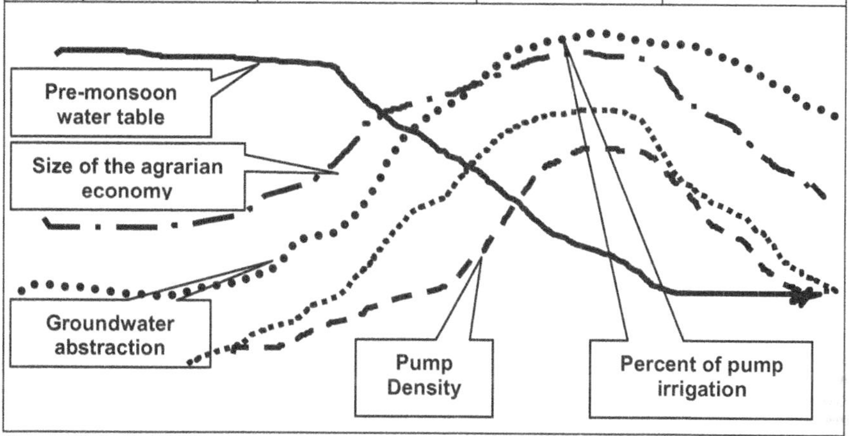

Source: Shah *et al*. 2003.

Figure 5.3 South Asia: Rise and fall of groundwater socio-ecologies.

5.2.4 Implications for IWMI research in Asia and Africa

Global discussions on groundwater irrigation in Asia are dominated by ecological concerns, totally ignoring its huge impact on livelihoods, incomes, poverty and productivity (Postel 1999; Seckler *et al.* 1998b). Occasionally, researchers have tried to bring a sense of balance to this discussion (e.g., Custodio and Llamas 2003). IWMI work has analyzed groundwater development as a process of constant interaction between socioeconomic and environmental variables; and it has focused on evolving practical approaches of managing this process.

One immediate impact of this macro-level assessment has been within IWMI itself. Startled by the meteoric rise in pumps in India, Pakistan and Bangladesh, IWMI scientists have explored whether similar pump expansion is imminent elsewhere. In Sri Lanka, government believed the total number of agro-wells to be around 30,000 by 2000; but a study by Kikuchi *et al.* (2003) estimated this number to be over 50,000 and pumps at 100,000. According to Barker and Molle (2003), in Vietnam, the rise in pump numbers has been 4.5 times over the

1990s, although it is not clear how many are used for pumping groundwater. Molle *et al.* (2003) assert that over the mainstream thinking dominated by large-scale, centrally managed irrigation systems, this groundswell of pumps "has superimposed a logic of individual, flexible, and on-demand access to water, which has far-reaching and, as yet overlooked, implications for the regulation and management of [our] water resources".

Table 5.2 South Asia: Rise and fall of groundwater socio-ecologies.

Feature	Stage 1 The rise of Green Revolution and tube well technologies	Stage 2 Groundwater-based agrarian boom	Stage 3 Early symptoms Groundwater overdraft	Stage 4 Decline of the Groundwater socio-ecology with immiserizing impacts.
Examples	North Bengal and North Bihar, Nepal terai, Orissa	Eastern Uttar Pradesh Western Godavari, Central and South Gujarat	Haryana, Punjab, Western Uttar Pradesh, Central Tamilnadu	North Gujarat, Coastal Tamilnadu, Coastal Saurashtra, southern Rajasthan
Characteristics	Subsistence agriculture; Protective irrigation Traditional crops; concentrated rural poverty; traditional water lifting devices using human and animal power	Skewed ownership of tube wells; access to pump irrigation prized; rise of primitive pump irrigation 'exchange' institutions. Decline of traditional water lifting technologies; rapid growth in agrarian income and employment	Crop diversification; permanent decline in water tables. The groundwater-based 'bubble economy' continues booming; But tensions between economy and ecology surface as pumping costs soar and water market become oppressive; private and social costs of ground-water use part ways.	The 'bubble' bursts; agricultural growth declines; pauperization of the poor is accompanied by depopulation of entire clusters of villages. Water quality problems assume serious proportions; some farmers begin moving out long before the crisis deepens; the poor get hit the hardest.
Interventions	Targeted subsidy on pump capital; public tube well programs; electricity subsidies and flat tariff	Subsidies continue. Institutional credit for wells and pumps. Donors augment resources for pump capital; NGOs promote small farmer irrigation as a livelihood program	Subsidies, credit, donor and NGO support continue apace; licensing, siting norms and zoning system are created but are weakly enforced. Groundwater irrigators emerge as a huge, powerful vote-bank that political leaders cannot ignore.	Subsidies, credit and donor support reluctantly go; NGOs, donors assume conservationist posture; zoning restrictions begin to get enforced with frequent political relaxations; water imports begin for domestic needs; variety ameliorative action starts.

Source: Shah *et al.* 2003.

Table 5.3 Groundwater governance challenges across hydro-geological settings.

Hydro-geological settings		Socioeconomic and management challenges			
		Resource depletion	Optimizing conjunctive use	Secondary salinization	Natural groundwater quality concerns
Major alluvial plains	Arid	●●	●o	●●●	●
	Humid	●	●●●		●●
Coastal plains		●●	●	●●o	●
Inter-montane valleys		●	●●	o	●
Hard-rock areas		●●	o	●	●●●

It is commonly believed that groundwater use in African agriculture is negligible. This may be true so far; but an IWMI study of the Dendron aquifer in Northern Province in South Africa showed that although the yields are generally low, the aquifer has been the sole source of irrigation water for commercial agriculture for more than 20 years (Masiyandima *et al.* 2002). Similarly, an IWMI-commissioned desk study of groundwater situation in the Limpopo Basin recently suggested that groundwater use in many parts of Africa might already be rising steeply. The Limpopo has only 35,000-odd irrigation wells at present; but 20 new ones are being made everyday; and at this rate, the number of irrigation wells will double by 2008, repeating the Sri Lankan and Vietnamese cases above. Though it supplies only 15% of total water, groundwater serves two-thirds of the Limpopo's people's needs. Smallholders are apparently 100% dependent on boreholes for whatever irrigation they manage to get (Tewari 2003). If this is true for other river basins as well, there is a powerful 'groundwater and poverty' dimension even in Africa. Tunisia has experienced an increase in irrigation tube wells from 60,000 in 1980 to 123,000 in 1997; and these supply 60% of its irrigation water (Bahri 2002). Morocco too has experienced rapid growth in groundwater irrigation (Turral pers. comm. Colombo, 2004); and initial IWMI estimates of groundwater-irrigated areas in Sub-Saharan Africa as a whole are nearly 50% higher than current official statistics (Giordano 2006).

It is clear that Africa cannot have a run on groundwater like many Asian countries because it just does not have the resource; but there already is a lot more groundwater irrigation going on in Africa than most people believe. Where groundwater is used intensively, the resource is declining and insufficient to meet future needs. Boreholes are difficult to site; and yields are low. Moreover, there are unfavorable economics due to high drilling costs. All in all, whereas many parts of Africa will see a rise in groundwater use, limits to sustainable use will be reached much earlier than is the case with much of Asia. IWMI will be in a better position to offer an analysis of groundwater possibilities in Africa after its socio-ecology study for that region is completed.

5.3 TECHNOLOGIES AND INSTITUTIONS

With a clearer understanding of the 'big picture' of groundwater irrigation, SGM research has worked to identify and promote research on promising technologies and institutions to enhance the sustainable use of the resource. Accordingly, this second SGM subtheme has concentrated on strategies to utilize groundwater irrigation as a means for poverty reduction; approaches to managing groundwater supply and demand; and the relationship between groundwater quality, agricultural production and public health. In addition, the subtheme has addressed a set of crosscutting issues related to management transfer of public tube wells and pump irrigation markets.

5.3.1 Groundwater for poverty reduction

Some 400 million of the world's poor are packed in eastern parts of the Ganga-Brahmaputra-Meghna Basin that encompasses eastern India, Nepal terai and Bangladesh-a region beset by small landholdings, low literacy rates, high levels of social stratification, feudal agrarian institutions, proneness to floods and waterlogging, and general economic stagnation. One of the few natural resources that the region has in plenty is groundwater; and this has remained under-developed because of chronic rural poverty that keeps people from investing in boreholes and pumps (Shah 2003b; Barker and Molle 2002; Kahnert and Levine 1993). The SGM theme emphasis in this region has been to explore intensive use of groundwater as a strategy for poverty reduction. According to IWMI research, manual irrigation technologies, such as treadle pumps, are particularly suited in this region especially for marginal farmers who can raise their incomes by up to $100/year by adopting treadle pump irrigation with a capital investment of $18-25 (Shah *et al.* 2000).

Further, IWMI studies have shown that because groundwater irrigation allows better control over the timing and quantum of water application to crops, access to the resource can help improve crop productivity, land use intensity and cropping patterns (Ballabh and Chowdhury 2003; Mukherji and Kishore 2003; Shah and Singh 2004). While governments - especially political leaders at national and regional levels - have recognized the critical importance of accelerated groundwater development, the SGM theme work has shown that:

(a) Existing programs for providing smallholder irrigation through government-managed tube wells have been a resounding failure;

governments should get out of these programs and instead hand existing government tube wells to farmer groups along the model developed and used effectively by the Government of Gujarat (Mukherji and Kishore 2003).

(b) Existing programs in India to provide small farmers subsidies to acquire pumps and boreholes can work better if redesigned as suggested by Shah (2003b) who showed that a powerful and virtuous 'dealer dynamic' was responsible for the acquisition of over 800,000 small pumps and boreholes by the poor, which belatedly launched the Green Revolution in eastern Uttar Pradesh.

(c) The best solution may well be to do away with subsidies altogether and open up imports of pumps; because of government subsidies, Indian pump prices are above those prevailing in either Pakistan or Bangladesh, none of which provide pump subsidies and both of which allow free import of cheap Chinese pumps (Shah 2003b; Shah, Hussain and Rehman 2000).

(d) Finally, development of the pump rental market can give a strong fillip to groundwater intensification and agrarian growth in this region; however, prospects for this are dimmed due to rural 'de-electrification' of the entire eastern portion of India since 1980.

5.3.2 Managing groundwater demand and supply

Groundwater policies in Asia have been far more successful in stimulating groundwater development and use for poverty reduction in groundwater-abundant regions than in regulating excessive groundwater draft where resource depletion has emerged as a major socio-ecological threat. Here, SGM has worked on demand and supply management strategies that might work in the Asian context such as (a) direct pricing; (b) indirect pricing; (c) promotion of water-saving technologies; and (d) the pros and cons of supply-augmentation through decentralized groundwater recharge. These strategies were implemented in 30 villages in India under IWMI's North Gujarat Sustainable Groundwater Initiative (see Box 5.1).

5.3.2.1 Pricing

Direct pricing is one demand-management strategy that has worked in several places in the industrialized world. In the western US, groundwater districts commonly levy a groundwater price based on volumes extracted by farmers. In the Murray-Darling Basin in Australia, states are assigned fixed salt credits, which regulate the amount of salts they are allowed to release into the Murray River. States, in turn, are planning to charge farmers based on saline water released by them. Under the new South African water law, 'user-pays, polluter pays' is enshrined in the national water regulatory framework. This marks some beginning to the use of economic incentives as an instrument of resource management. In the Asian setting too this might work; but the question is of the numbers involved on the one hand and the priorities as well as capacities of the groundwater institutions, on the other.

Box 5.1 North Gujarat initiative: A case study in groundwater management.

Applying the lessons from its research on managing groundwater irrigation in a sustainable manner, IWMI commenced the North Gujarat Initiative. In many regions suffering from groundwater depletion, reversing farmer fatalism is a key aspect of effective intervention. In order to explore ways of restoring sustainable farming approaches in such ecologies, IWMI's North Gujarat Sustainable Groundwater Management Initiative (NGI) has begun working with irrigators in 30 villages in Banaskantha district in four areas:

(a) promotion of micro-irrigation and water-saving irrigation practices in crops like alfalfa;
(b) promotion of water-saving crop alternatives;
(c) decentralized groundwater recharge activities; and
(d) proactive water education of farmers, women and school children. During its 2 years of operation, NGI villages have become a kind of laboratory to experiment with a variety of technologies and sustainability approaches. Key results are as follows:

Firstly, drip irrigation in alfalfa was found to reduce water application in the range of 7.3%–43% and to increase yield/ha in the range of 7.3%–10.8%, but economic returns from drip irrigation were not very attractive with the benefit/cost (B/C) ratio ranging from 1.09 to 1.29. However, when true costs of energy used in pumping are built into the analysis, the B/C ratio improved to 1.18–1.83; in addition, when water is valued at opportunity cost—reflected in price charged in local irrigation water trade—the B/C ratio improved further to 1.28–2.78 (Kumar *et al.* 2003);
Secondly, a comparative study of irrigation water productivity of water-selling well owners, water buyers and shareholders of tube well cooperatives showed that groundwater buyers achieve higher water productivity compared to well owners whose marginal cost of water is artificially lowered due to flat electricity tariff and are therefore not confronted with the true marginal cost of pumping. Similarly, shareholders of tube well cooperatives who face rationed water supplies achieve higher water productivity by choosing water-saving crops (Kumar 2005);
Finally, a study of groundwater depletion in the Sabarmati River Basin concluded that with alarming drops in groundwater levels and reduction in well yields, well irrigation in the basin has increasingly become economically unviable, though farmers are able to continue irrigation with deep tube well pumping due to heavy electricity subsidies (Kumar *et al.* 2002).

In India, Pakistan and Bangladesh, there are already over 21 million private tube well owners who are neither registered nor licensed. Locating and registering these, installing meters on tube wells, reading meters at regular intervals and recovering a volumetric price may pose a formidable logistical challenge. As a result, there is hardly any discussion of water pricing as a demand-management option.

The realm of possibility expands rapidly as economies grow, as IWMI's more recent SGM studies suggest. With comparably high population density, pressure on irrigated agriculture and high tube well density, Chinese water administration looks far more keen and better prepared-compared to the South Asian ones-to enforce some kind of rule of law in its groundwater economy (Shah, Giordano and Wang 2004). The Chinese have already begun to enforce tube well permits on agricultural tube wells and water withdrawal permits and water resources fees on urban and industrial users of groundwater. In general, comparing with South Asia suggests that many instruments of resource management that appear impractical at South Asia's current stage of development may not be so when their economies have grown.

5.3.2.2 Energy pricing as a surrogate

Since groundwater withdrawal requires energy, pricing of energy can be an indirect means to price groundwater. Throughout South Asia, the nexus between energy and groundwater irrigation has acquired huge dimensions; and under the SGM theme, IWMI has focused much energy and effort on this nexus. This topic is discussed in section 5.4.3; but it is relevant to note here that IWMI's work in China does suggest that energy cost has the potential to act as a groundwater demand-management tool (see Figure 5.4). To an increasing extent, energy prices are also beginning to act as a surrogate for groundwater price in Pakistan and Bangladesh as we shall discuss later. In the North China Plains where electricity subsidies to agriculture are absent or marginal, rising energy costs of pumping groundwater is an important driver of growing adoption of green houses, sprinklers, drip irrigation and high-value crops (Shah 2003a). Shah, Giordano and Wang (2004) suggest that falling rice prices and high energy costs of pumping have formed a pincer that is compelling farmers to take to groundwater-saving cropping patterns and technologies in many parts of North China Plains. Based on field visits, they report growing areas of drip-irrigated genetically modified (Bt) cotton, replacing the maize-wheat cycle in Shandong and Hebei; and rain-fed maize replacing rice in Liaoning.

5.3.2.3 Promotion of micro-irrigation technologies

As a tool for groundwater demand management as well as for energy conservation, the SGM theme has also explored the potential for propagating micro-irrigation technologies such as drip and sprinkler systems (Shah and Keller 2002). While there is growing consensus that micro-irrigation technologies increase the productivity of water applied, doubts remain about whether it can result in net reduction in groundwater use at the basin level. First, in closed basins, it is argued that groundwater *not* pumped is available for other uses that would claim it. Second, drip irrigating farmers use up water saved by expanding irrigated areas or land use intensity. As a result, micro-irrigation may produce local gains but it is a zero-sum-game at the basin level. IWMI is presently undertaking studies to explore these issues in North Gujarat, where it is operating an action research project in 30 villages, and in the Maikaal region of Central India where it is studying farmers "growing organic cotton under groundwater stress" (Shah and Keller 2002; Verma *et al.* 2004).

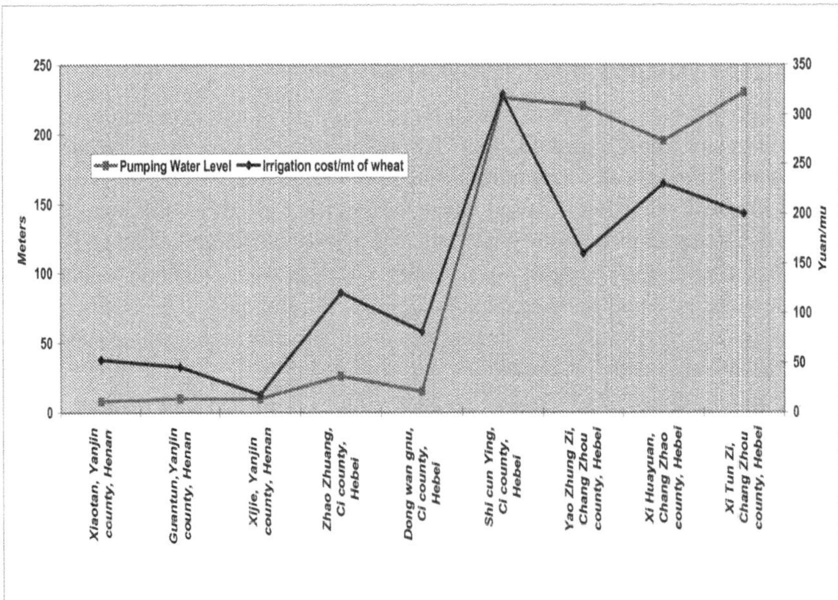

Figure 5.4 Relationship between irrigation cost and levels of water pumping in North China Plains.

IWMI researchers found other uses for micro-irrigation too. In close collaboration with the International Development Enterprises (IDE), IWMI researchers explored the livelihood-boosting potential for land-poor rural households (Shah and Keller 2002; Verma 2003, 2004; Upadhyay *et al.* 2005). They also found that farmers under severe groundwater stress-such as cotton growers in the Maikaal region in Central India-reach out spontaneously for such contraptions as a coping mechanism, provided these are available off the shelf in their surrounding area at an affordable price. Efforts are now underway to cut costs of micro-irrigation systems and to redefine the product concept from custom-solution to 'package solution'. A farmer-invented *pepsee* system in India, for example, costing all of $100/acre (as compared to $1,200/acre for the simplest buried strip drip system in the US) has now become the basis for IDE's new Easydrip system which offers much better quality at around $350/acre (Verma *et al.* 2004).

5.3.2.4 Conjunctive use of surface water and groundwater

An area in which IWMI has a long tradition of work is in the conjunctive use of surface water and groundwater, especially in canal commands.[4] IWMI's conjunctive water use work in Pakistan and Indian Punjab has a strong focus on managing secondary salinity. Some of the key ideas and results of this work are summarized in section 5.3.3. However, some research on conjunctive management has also focused on supply-side management strategies where secondary salinization is not a primary issue. Some notable work resulted from a 10-year long collaboration between IWMI and India's Central Soil Salinity Research Institute in Madhya Ganga project, which highlighted how monsoonal flood waters stored in unlined irrigation canals can be used to transform the groundwater hydrology of a region for optimal control of water level and for reducing energy costs of pumping (IWMI-Tata 2003a).

5.3.2.5 Decentralized recharge as a mass movement

IWMI work has also explored decentralized groundwater recharge by village communities on a massive scale in drought-prone regions of western Rajasthan and the Saurashtra region of Gujarat where water has become 'everybody's business' (Agarwal *et al.* 2001). IWMI began studying this movement by trying to understand what catalyzed it first of all and then kept energizing it (Shah 2000). The key conclusion was that while some religious leaders and their organizations played a crucial role in pioneering recharge through modifications in irrigation wells and construction of other community water-harvesting

[4] One of the most closely researched areas is the Rechna Doab in Pakistan's Punjab, where IWMI studies in conjunctive use and management have been in progress since 1985. An early review of IWMI work here can be found in Rehman *et al.* 1997 and Rehman 1997.

structures, the movement gathered momentum because of the growing evidence of the usefulness of such structures for combating drought and moisture stress (Shah 2000). IWMI has since explored four new questions: (a) In principle at least, can decentralized water harvesting and groundwater recharge result in net improvements in basin- or region-level welfare? (b) If not, is there emerging evidence of the movement waning now that it has operated in a hyper-active mode for over a dozen years? (c) Has the decentralized water-harvesting and recharge movement stayed just that - a movement - or has it marked the first step to decentralize water resources management by communities? and (d) For populous, water-scarce countries like India, does Saurashtra represent an exception or the harbinger of a broader, mainstream trend?

The evidence collected by IWMI and its collaborators (Shingi and Asopa 2002; Sakthivadivel and Nagar 2003; Nagar 2002; Shah and Desai 2002) offers some tentative answers:

(a) Decentralized groundwater recharge can at least ensure security of the main *kharif* (rainy season) crop for most farmers in Saurashtra and Kutch (in Gujarat); and there seems no significant sign yet of any large opposition to water harvesting in the catchment areas of river basins (Shingi and Asopa 2002; Shah and Desai 2002).

(b) There seems little evidence of the waning of people's faith in the power of decentralized water harvesting to improve their livelihoods.

(c) There are some early signs of an emerging consciousness of the need for water-demand management, especially in agriculture; but this is essentially in response to the need to save crops from declining well yield (Shah and Desai 2002).

There is no clear answer to the last question since Saurashtra and Kutch are different from other parts of Gujarat and western India in several aspects of their socio-ecology.

There is evidence that the recharge movement has produced broad-based positive impacts (Shingi and Asopa 2002). The primary benefit is ensuring the security of the kharif crop, which farmers in Saurashtra and Kutch are unsure of in 3 out of 5 years because of frequent early withdrawal of monsoons. The water harvested and available close to the point of use has ensured that the kharif crop is saved from moisture stress towards the close of the season; and the social value of this benefit seems indeed to be great. This is enough to induce farmers to take farming seriously again and to invest in land care and inputs (Shingi and Asopa 2002).

While many critics believe that impacts of water harvesting and recharge movement are localized, research suggests that the impacts are more widespread. For example, an IWMI-sponsored study compared the impact of the recharge movement on the rise in static water level (SWL) during the monsoon using pre-monsoon and post-monsoon records of groundwater levels in 30 *talukas* (subdivision of a district) maintained by the Gujarat Water Resources Development Corporation. The results of the study are summarized in Figure 5.5. The first pair of maps shows the long-term average in rainfall and SWL, respectively; and the next two pairs show the rainfall and SWL changes during the monsoon during 2000 and 2001, respectively. The long-term average maps show lower monsoonal SWL rise despite higher rainfall. The latter maps show that during 2000 and 2001, although rainfall has been less than average, groundwater availability at the end of the monsoon is better in most talukas.

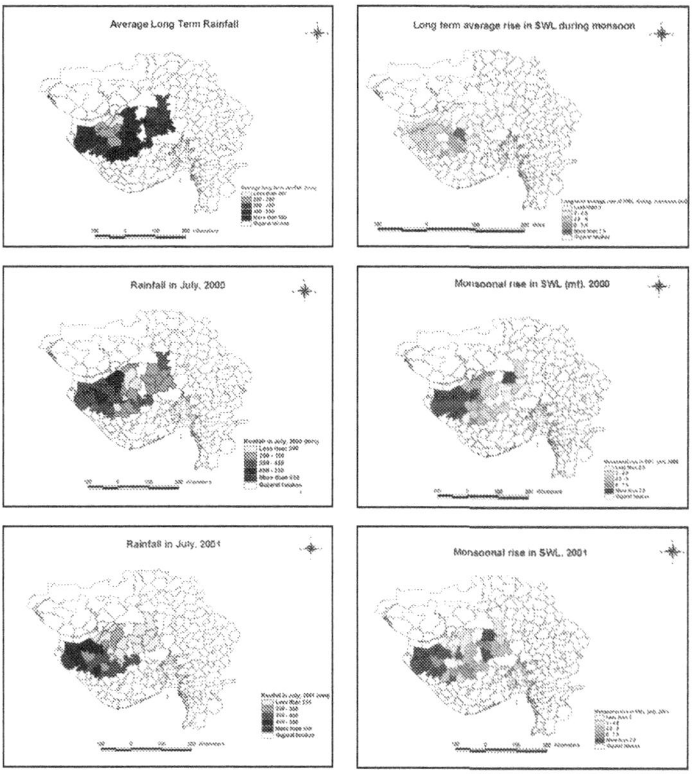

Figure 5.5 Impact of recharge movement on groundwater level, Saurashtra, Gujarat, India, 2000-01.

5.3.3 Coping with poor-quality water

SGM researchers have focused on two aspects of groundwater quality. First, from a public health standpoint, IWMI has explored the risks of arsenic and fluoride contamination in parts of South Asia as well as possible remediation options. Secondly, IWMI has explored approaches to manage primary and secondary salinization. Both research topics are summarized below.

5.3.3.1 Groundwater quality and public health

No country in Asia has come to depend on groundwater for food and livelihoods as much as Bangladesh, which due to flood risks, could never rely on surface irrigation. From 4% in 1972, the share of groundwater in Bangladesh's irrigated area has soared to over 70% in 1999; as a result, its rice production grew from 9.8 m mt (million metric tons) to 20 m mt over that period as yield/ha jumped from 1.05 m mt/ha to 1.97 m mt; and the cropping intensity shot up from 145% in 1975 to 175% in 1999, made possible by increased shallow tube well irrigation of *boro* rice (Mainuddin 2002).

This overindulgence in groundwater irrigation has recently begun to show a darker side. Growing evidence suggests that Bangladesh - and the neighboring Indian state of West Bengal - may be facing a public-health crisis, on account of the presence of arsenic in groundwater used by most people for drinking. Groundwater sources in 61 out of Bangladesh's 64 districts are contaminated with arsenic; and an estimated 35 million people are at risk of being exposed to arsenic poisoning through drinking water. IWMI conducted an initial study on the arsenic crisis in 1999 by carrying out a literature-based situation analysis (Raschid-Sally 2000).

In regions of South Asia not affected by arsenic, fluoride contamination of groundwater is rapidly emerging as a public-health issue. During 2003, IWMI initiated some exploratory work on assessing the prevalence rates of dental and skeletal fluorosis in North Gujarat and southern Rajasthan, regions where high fluoride in groundwater has been known for over two decades. In North Gujarat, IWMI surveyed 28,000 people from 25 fluoride-affected villages selected from Patan and Mehsana districts; similarly, 6,600 people were interviewed in Banswara and Dungarpur districts of southern Rajasthan. The results - shown in Figure 5.6 - are a cause for concern. These show that: (a) the prevalence of dental fluorosis is significantly higher than that of skeletal fluorosis; (b) prevalence rates of all symptoms of fluorosis increase with age; (c) women show higher prevalence rates compared to men; (d) in the age group 46-60 years and above 60 years, 8-15% of the people interviewed had become crippled; and (e) our sample from southern Rajasthan-consisting of *Bhils* (poor tribal people) -

showed higher prevalence rates starting from a younger age compared to our sample in North Gujarat.

Much research and experimentation is underway worldwide on arsenic as well as fluoride. A lot of it has to do with evolving low-cost technologies for removing the contaminant from drinking water supplies at household or community level. During the past 3-5 years, a range of filters has been on the market. Western Indian states have undergone a generation of community defluoridation plants; but nowhere has the experience been encouraging. Following a growing trend in North Gujarat towns for private entrepreneurs investing in and operating Reverse Osmosis (RO) plants on a commercial basis, the IWMI-Tata Program supported two experimental plants at village level which are doing well suggesting a possible option. Several other household-level options are also available - such as *Mytree* being promoted in Andhra Pradesh for fluoride (using alum) and a dozen arsenic filters being popularized in Bangladesh. However, while it has been common knowledge for over 30 years that alum can separate fluoride, and alum-based fluoride filters have been distributed free for an equally long period, the adoption of these technologies is poor. It is therefore IWMI's aim to understand why and to help generate a better understanding of factors that promote the uptake of the various filtration options.

5.3.3.2 Conjunctive use in saline areas

In dry alluvial plains of the Indus and western Gangetic Basin, intensification of groundwater irrigation over the past five decades has resulted in aquifer depletion and the upcoming of the fresh-saline water interface. Equally, over-irrigation with surface water has caused waterlogging and salinization. Irrigation with marginal to poor-quality groundwater has added to soil salinization and sodification. In IWMI's SGM work in Pakistan and Indian Punjab, Sindh, Haryana, a basic premise has been that all these problems can be addressed through effective conjunctive management of surface water and groundwater (Qureshi *et al.* 2002a; Qureshi and Masih 2002b).

Three distinct approaches have been explored. First, in an ACIAR-supported project, conjunctive management strategies - focusing on technical and institutional approaches - were explored in Rechna Doab in Pakistan Punjab and the Murrumbidgee Irrigation Area (MIA) of Australia. MODFLOW and MT3D simulation models were used to generate sustainable levels of groundwater and surface water use at regional scales and also to test the efficacy of the subsurface evaporation basin for drainage management at various levels. Resistivity surveys identify potential sites in the Rechna Doab for inducing artificial recharge. The SWAGMAN-farm model was also used to test the financial and economic viability of technical solutions generated (Christen and Van Meerveld

2000; Jehangir and Horinkova 2002; Khan *et al.* 2003); their general conclusion was that appropriate institutional arrangements for conjunctive management of groundwater and surface water are needed to make such solutions work.

(a) North Gujarat

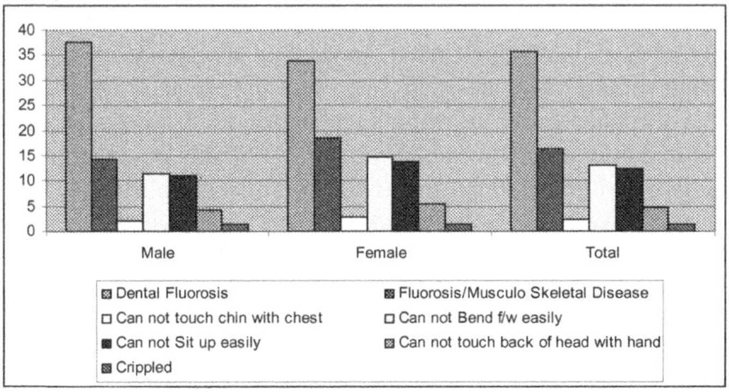

(b) Rajasthan

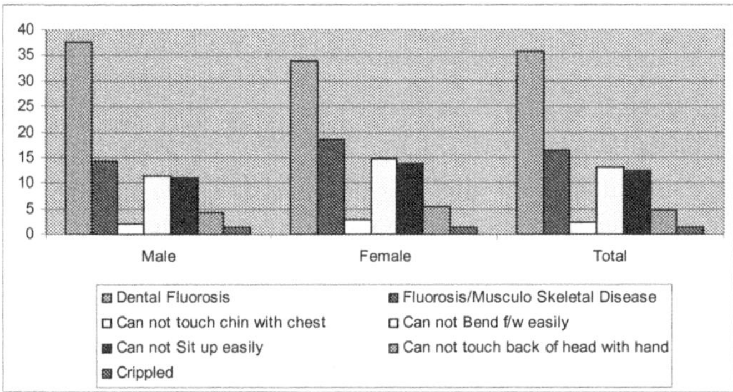

Figure 5.6 Fluoride impact in North Gujarat and Rajasthan, India.

The second approach - tried in a project in Pakistan Punjab with funding from the National Drainage Programme (NDP) - identified and tested promising skimming-well technologies to control the saline groundwater intrusion as a consequence of pumping. In the dry alluvial plains of South Asia as well as parts of North China, the native groundwater is deep and saline because of the marine origin of the hydro-geologic formation. Percolation of freshwater has formed a fresh groundwater lens above the underlying native saline groundwater layer. In saline groundwater areas, which cover 22% of Punjab, and 78% in Sindh, fresh groundwater lenses are thin (approximately 35-40 meters as compared to 150 meters in fresh groundwater areas); but according to an estimate, these probably store about 200 billion m^3 of freshwater. Already, farmers are after this water and have tried to tap through nearly 10,000 multi-strainer shallow tube wells having small bores. However, most of the wells, especially installed in the saline groundwater areas, are extracting groundwater from inappropriate depths, at improper discharge rates, and above all, they follow inadequate operational schedules. As a result, there are problems of deteriorating quality of pumped water, and large numbers of wells have already been abandoned due to this reason. Clearly, there is a role for good science here; hence, IWMI's work on skimming technologies (Asghar *et al.* 2002a, b; Saeed *et al.* 2003).

A third approach deals with secondary salinization in dry alluvial plains. In these regions, every time groundwater is used for crop production, plants extract the freshwater and leave behind a more saline fluid in the soil. If salts are not leached, either using additional freshwater or through rainfall, salts build up in the soil and affect plant growth. Even with leaching, salts enter the top layer of the shallow groundwater.

This mobilization and vertical recycling will ultimately reduce soil and groundwater quality. Numerous experiments of improved irrigation practices using water of different quality and for reclaiming salt-affected soils, mostly at field level, have been undertaken but with limited success.

IWMI's activity focused on upscaling field-scale knowledge to large irrigation systems to enhance environmental sustainability of irrigated agriculture. Long–term impacts of using different-quality groundwater in isolation and in conjunction with canal water in different proportions on crops and soil salinity were evaluated for the wheat-cotton agro-ecological zone of Punjab, Pakistan.

The Soil-Water-Atmosphere-Plant (SWAP) model was used to simulate different scenarios using 15 years' actual climatic data. The results showed that even in fresh groundwater areas (EC = 1.0 dS/m), where the risk of using groundwater alone for irrigation will be relatively limited, occasional leaching of salts with the canal water will be helpful to maintain long-term sustainability of irrigated agriculture. In marginal groundwater areas (EC = 1.5 dS/m), mixing groundwater and canal water in the ratio of 1:1 will keep the soil salinity within acceptable limits. However, additional leaching with freshwater would be needed in relatively dry years, as the risk of soil salinization is higher. In saline groundwater areas, irrigation with saline groundwater (EC = 3.0 dS/m) will be a complete disaster.

Mixing this groundwater with canal water will not help in reducing the risk of soil salinization. The results further indicate that present farmer practices of applying more frequent irrigation with saline water do not help in removing salts from the root-zone. In such areas, only a complete change in the farming system itself can improve the situation and support more sustainable means of production (Qureshi and Masih 2002a, b; Qureshi *et al.* 2002b; Sarwar and Bastiaanssen 2001). Bringing about such sweeping change is not a question just of technological choice but also of dealing with a host of existing and new institutional constraints (Jehangir and Hornikova 2002; Jehangir *et al.* 2002).

5.3.4 Cross-cutting issues

SGM research has also been carried out on alternative institutional arrangements from the viewpoint of their impacts on productivity, equity and sustainability of groundwater irrigation. Throughout the 1950s and 1960s, governments in India, Pakistan and Bangladesh established tube well programs owned and managed by the government bureaucracy primarily to ensure equitable access to groundwater irrigation. Evaluations carried out as far back as in the 1970s had begun hinting at their poor overall performance; and by the 1980s, new investments in public tube wells had begun to taper off. However, an issue many governments faced was about what to do with existing tube wells in the government sector. Under bureaucratic management the cost of operating these has been so high - and their operating efficiency so low - that public tube well programs everywhere in South Asia have been losing money on a cash basis. The issue was not so much the technology - although Palmer-Jones (1995) has argued strongly that the deep tube well technology popularized by donors is ill suited to South Asia - but rather the management and crafting of appropriate local institutional arrangements that ensured viable operation. A study by Mukherji and Kishore (2003) suggested that Gujarat's turnover of public tube wells to farmer groups may offer a good model to many governments saddled with loss-making public tube well programs. Two key elements of Gujarat's program that explained its success were: a proactive thrust towards management transfer by the government managers; and a flexible, eclectic view of what constituted a beneficiary organization. A somewhat similar model has also been found in China (Shah, Giordano and Wang 2004).

As institutions that expand equitable access to pump irrigation, groundwater markets have been of crucial interest to the SGM theme work.[5] In early South Asian studies on fragmented pump irrigation markets, the extent of market development was gauged in terms of its 'depth' and 'breadth' (Shah 1993). Market 'depth' was typically measured as the proportion of water output sold by pump irrigators in a given year or season and purchased water as the proportion

[5] A number of IWMI studies have been conducted on this topic. See, for example, Palmer-Jones 1994; Strosser and Kuper 1994; Shah, Hussain and Rehman 2003; Deb Roy and Shah 2003; Shah *et al.* 2006.

of water used by nonowners, 'breadth' of the market was measured by (a) the proportion of pump owners in a village who sold pump irrigation; and (b) the proportion of non-pump owners who purchased pump irrigation. One of the key institutional insights from the five-country survey of Groundwater Socio-ecology was the change in the 'depth' and 'breadth' of pump irrigation markets in different parts of South Asia.

During the late 1980s, researchers found pump irrigation markets to be very deep and broad in western and peninsular India whereas they were shallow and narrow in eastern India. By the turn of the millennium, IWMI research showed that the situation seemed ripe for reversal (i.e., that pump irrigation markets are becoming thin and shallow in western and peninsular India as well as in Pakistan Punjab, whereas they have greatly increased in depth as well as breadth in eastern India, Nepal terai and Bangladesh).

Several explanations are possible but two seem particularly relevant:

(a) First, pump irrigation markets thrive only where pump densities are low but irrigation demand is high. Western and peninsular India and Pakistan Punjab - which began their tube well revolutions early - have achieved high levels of pump density. As more farmers acquire their own pumps and bores, they opt out of the market. But in the eastern Indo-Gangetic Basin (IGB), pump densities have begun rising only recently, which explains the region's vibrant pump irrigation markets.

(b) Second, in many regions of western and peninsular India, which suffer from a high level of groundwater stress, pump owners are opting out of the market because they can barely meet their own irrigation requirements. Elsewhere, monopoly rents on pump irrigation are rising, suggesting that prevailing prices reflect not only pumping costs and scarcity rent of pumps but also scarcity rent on groundwater itself (Dubash 2002; Janakarajan 1994; Kolavalli and Chicoine 1989; Palmer-Jones and Mandal 1987; Shah 1993).

However, deep and broad water markets in the eastern IGB may not necessarily be 'efficient'; and these efficiency and equity impacts of pump irrigation markets, which were a subject of much discussion during late 1980s, can be seen in full play in eastern IGB. Roy and Mainuddin (2003), who surveyed pump irrigation sellers and buyers in 40 villages of Bangladesh, found that farmer-financed shallow tube wells and diesel pumps (80% of all pumps) dominate the pump irrigation economy. Of pump owners interviewed, 87% sold water, indicating broad markets. On average, a tube well owner sells water worth Tk 24,700/yr. (approximately, $425). Projecting for the country as a whole, the authors estimated the revenue from pump irrigation sale for

Bangladesh at Tk17.6 billion (approximately $303.5 million). With high pump density, Bangladesh's pump irrigation markets are highly competitive; however, a critical problem facing groundwater irrigators is the high-energy cost. Both diesel and electric energy prices are high and rising steadily, posing the threat of pricing the buyers out of pump irrigation markets. Much the same scene is evident in Pakistan Punjab and Sind where metering of tube wells in 2000 and imposition of electricity tariff at near-commercial rates have begun to alter the dynamic of pump irrigation markets.

Have groundwater institutions evolved differently in North China compared to South Asia? Preliminary results show that they have (Xiang *et al.* 2000). Based on research in Hebei province, the authors found that since the implementation of the Household Responsibility System in the early 1980s, China's increasingly aged, damaged and poorly maintained rural water projects have become the main constraints to agricultural growth. In order to ensure agricultural development, the system of property rights within the groundwater irrigation system has been gradually innovated, offering different levels of private and collective rights. Using econometric analysis of data from 30 villages within three counties in the Hebei province, this study showed that property rights innovation-in particular, increased levels of 'non-collective' property rights-has led to the expansion in cultivated areas under cash crops and an increase in farmers' income. Besides property rights innovation, government grain procurement policy, market prices of agricultural products and the opportunity cost of agricultural labor are all important factors that influence farmers' behavior in the adjustment of cropping patterns.

5.4 GOVERNANCE FOR SUSTAINABILITY

Global discussion on institutions and policies for sustainable groundwater management has focused heavily on demand management through one or a combination of the following approaches:

(a) *Direct management* by groundwater agencies through (i) legal and regulatory interventions and their administrative enforcement; (ii) creating new institutions (such as water markets) and modifying property rights on water, including creating tradable rights (Rosegrant and Gazmuri 1994; Thobani 1995; Bauer 1997, 1998; Meinzen-Dick and Mendonza 1996; Government of India 1999); and (iii) the use of economic incentives such as water resources fees. All these are being used with varying levels of success in several countries, but SGM has followed the experience of Mexico and China (Shah 2003a; Shah, Giordano and Wang 2004). An extensive overview of some of the direct management strategies that have been tried out in industrialized countries is available in Turral 1998.

(b) *Community management*: some research groups have drawn attention to NGO experiments, such as Tarun Bharat Sangh's River Parliament in eastern Rajasthan, the Sukhomajari project in Haryana and Salunke's Pani Panchayats in Maharashtra, to argue that given the right conditions, local communities can be empowered to evolve mechanisms for sustainable resource management (British Geological Survey 2004; Shah 2004); community management of groundwater is also part of Mexico's national policy (Wester *et al.* 1999; Shah, Scott and Buechler 2004).

(c) *Adaptive approaches*: In a series of recent writings (Moench 1994b; Moench, Burke and Moench 2003; FAO 2003; Moench *et al.* 1999), researchers have developed and advocated as an 'expanded management perspective' adaptive, livelihoods-based approaches to groundwater management especially in countries like India. Conventional approaches do not address the livelihoods systems from which the structure of demands on groundwater emerges and which are taken as 'givens'. These would not be taken as givens if groundwater management were to lean more on livelihood-focused adaptive approaches that would more likely work.

(d) *Strategic indirect management:* SGM has argued that while direct management presents unique logistical challenges, there lie opportunities for indirect management of groundwater demand. Unfortunately, levers for large impact on groundwater demand are often not controlled by groundwater managers but by decision makers in other sectors. In India, for example, it is argued that annual groundwater withdrawal for agriculture in some of the worst overexploited areas may fall by 12–20 km^3 just by eliminating the perverse electricity subsidies. Likewise, the rice-wheat system of northwestern India-responsible for much groundwater depletion in Punjab, Haryana and Western UP - can be reconfigured within 5 years by reorienting India's food grain procurement policies towards eastern India while northwestern India shifts to cropping patterns that offer more cash per drop of groundwater. In many groundwater-stressed regions of India, old and new surface water structures such as canals, tanks and drains, coveted in the past for flow irrigation, are increasingly being used primarily as groundwater recharge structures. SGM has argued that rather than trying to preserve these as surface irrigation structures, there might be value in working to enhance their productivity as recharge structures.

5.4.1 Comparative institutional analyses

In exploring these four management alternatives, IWMI has engaged in a comparative study on groundwater management institutions and policies in South Asia, North China and Mexico, three regions of the world where agriculture, food and livelihoods depend heavily on increasingly unsustainable use of groundwater (Shah 2003a; Shah, Scott and Buechler 2004; Steenbergen and Shah 2003). Table 5.4 summarizes some key conclusions from this comparative analysis.

Table 5.4 Groundwater governance: Comparative analysis of institutions and policies in South Asia, China and Mexico.

No.	Particulars	South Asia	China	Mexico
1	Government share in GW provision to agriculture	Miniscule; <0.01%	No	No
2	2. State provision of GW to urban settlements	Significant	Significant	Significant
3	State participation in GW monitoring	Yes	Yes	Yes
4	Incentives to private investment in GW development	Significant in India and Sri Lanka, often perverse; discontinued in Pakistan, Nepal, Bangladesh	None or insignificant	None
5	Direct subsidies to tube well operating costs	Huge in India; less in other countries	Nil or insignificant	Yes, energy subsidies
6	Registration of GW structures	No	No	Yes
7	Permits to abstract GW	No	Yes, but mostly to villages, municipalities and industries	Yes, but water quantities unenforceable
8	Promotion of water-saving technologies	Ineffective	Yes, strong	Some
9	Promotion of small-scale water-harvesting and recharge works	Strong in western India; but growing elsewhere in hard-rock India	South-North water transfers	Yes, in highlands where *bordos* (small tanks) are the mainstay of livestock farmers

Our overriding impression is that South Asian countries have not even begun to address the problem in any serious manner. China has, but it will take time before its initiatives bear fruit. Mexico has gone the furthest in creating a legal and property rights structure that might be drawing a leaf from an institutional economics textbook. Interestingly, we find little evidence that these have helped Mexico move towards sustainability and it is early days yet; but Mexico's efforts will need to produce better results before they inspire other groundwater-using countries (Shah, Scott and Buechler 2004). Even Israel, the Mecca of water management-which has deployed the entire array of groundwater regulation instruments including centralized control, conjunctive use, metering and control of pumping-has been unable to check progressive depletion of its aquifers (Fietelson in IWMI 2003:19).

A key conclusion is: how countries respond to the challenge of sustainable management of their groundwater economies depends on a constellation of factors that define the unique context of each country. This constellation differs vastly across regions and countries; and these differences have a decisive impact on whether an approach that has worked in one country will work in another. As a simple illustration, Table 5.5 sets out some key variables that define the organization of the groundwater economy in six different countries, which make intensive use of groundwater in agriculture. For example, the table illustrates that the US uses around 100 km^3 of groundwater for irrigation; but to manage its economy, it has to monitor and regulate only around 200,000 pumping plants, each producing around 500,000 m^3 of groundwater/year. By comparison, India uses over 200 km^3; but to manage this groundwater economy, it has to manage the owners of over 20 million small wells, each producing an average of 8,000 m^3 of water/year.

The nature of the political system also matters. Iran has been able to impose a complete ban on sinking of new tube wells throughout its central plains that encompass two-thirds of the entire country (Hekmat 2002). But Mexico has been trying to ban new tube wells in its *bajio* for 50 years, and has not yet succeeded (Scott in IWMI 2003b:18). China has a large number of tube wells scattered over its vast northeastern countryside; yet chances are that over the coming decade, it will be able not only to bring these within the ambit of its permit system but also succeed in influencing their operation (Wang in IWMI 2003b: 16-17; Shah, Giordano and Wang 2004). Political structures and systems will hinder similar efforts in India for some time to come.

Besides what is feasible and practical, there is also the question of social impacts of approaches adopted. In Mexico and the US, where a miniscule proportion of people depend on groundwater for livelihoods, governments may adopt a tough regulatory posture more easily than in South Asia, where over half of the total population may directly or indirectly depend on groundwater use for their livelihood. Here, it is not surprising that political and administrative leadership is reluctant to even talk about regulating groundwater use, leave alone acting on it. Even in China, where political resistance from farmers is not a serious issue, and Mexico where the irrigators are small enough to be ignored, governments have steered clear of tough regulatory measures (Shah 2003a).

Table 5.5 Structure of national groundwater economies.

Country	Annual GW use (km^3)	No of GW structures (million)	Extraction/ Structure (m^3/year)	Percent of population dependent on GW
India (in 1993-94)	150	19	7900	55-60
Pakistan-Punjab (mid-1990s)	45	0.5	90000	60-65
China (mid-1990s)	75	3.5	21,500	22-25
Iran (1990)	29	0.5	58,000	12-18
Mexico (1990)	29	0.07	414,285	5-6
USA (1990)	100	0.2	500,000	<1-2

5.4.2 Strategic management versus direct regulation

If the world's water crisis is 'mainly a crisis of governance' (Global Water Partnership 2000), groundwater represents the grimmest side of this crisis in Asia. The Australian Groundwater School at Adelaide is apt in choosing a credo that says, "Groundwater will be the enduring gauge of this generation's intelligence in water and land management". However, just 5.5% of Australia's irrigated areas depend on groundwater; the corresponding proportion for China is 28%, for India 56-60% and for Bangladesh 90%. Further, at some 12-15 million, the number of individual tube well owners who pump groundwater for irrigation in the Indo-Gangetic and Yellow River Basins is probably 1,000 times what Australian groundwater managers will ever have to cope with. As a result, the issue of governing groundwater economies in Asia, especially, their agricultural component - looks difficult, unless population pressure on farmland declines substantially.

IWMI's thinking about the way to go in the short run is to look for strategic instruments of indirect groundwater management and leave direct regulation for an appropriate future date. A hardware-centric, supply-side strategy has been to import surface water to wean farmers away from groundwater, as Spain's plans to transfer water from north to south and China's project to transfer it south to north. Responding to groundwater depletion in western and southern India is at the heart of India's new river-linking project. Another is redesigning a region's agriculture as if water endowments mattered. Indian Punjab is facing severe groundwater stress from its rice-wheat economy; Amarinder Singh, the state's Chief Minister has figured, correctly in our view, that it will be easier to reconfigure Punjab's agriculture than to control groundwater extraction. He has therefore embarked upon weaning Punjab's farmers away from rice-wheat to value-added crops, hoping that from orchards, his farmers will make more money and use less water. The same logic can be extended to all of India: some 70-90 km^3 of groundwater is pumped in western and peninsular India to grow food grains that eastern India can easily produce, if only the government food procurement machinery were reoriented away from Punjab-Haryana-western Uttar Pradesh to Bihar-West Bengal-Orissa-eastern Uttar Pradesh-Assam, a region that does not quite know what to do with over 1,200 km^3 of flood water it receives every year. A possible conclusion is that the Food Corporation of India may be able to do more for groundwater governance in India than the Central Groundwater Authority. Agricultural input and output prices - and subsidies - play a big role in vast areas of Asia subject to groundwater overdraft.

5.4.3 Energy-irrigation nexus

Electrical utilities can play an important role in groundwater demand management. In South Asia, which uses energy worth US$5-6 billion/year (InRs 25,000 crores) to pump some 210 km^3 of water mostly for irrigation, little can be done in the groundwater economy that will not affect the energy economy; and the struggle to make the energy economy viable is frustrated by the often violent opposition from the farming community to efforts to rationalize energy prices. As a result, the region's groundwater economy has boomed by bleeding the energy economy. IWMI's SGM theme research suggests that this does not have to be so; and the first step to evolving approaches to sustaining a prosperous groundwater economy with a viable power sector is to manage energy and groundwater as a nexus (Shah, Scott *et al.* 2004). IWMI research suggests that the inability to manage groundwater and energy economies as a *nexus* is a great opportunity missed. In South Asia, there seems no practical means for direct management of groundwater; laws are unlikely to check the chaotic race to extract groundwater because of the logistical problems of regulating a large number of small, dispersed users; water pricing and/or property right reforms will also not work for the same reasons. Power supply and pricing policy offers a powerful toolkit for *indirect* management of both groundwater and energy use.

In particular, the SGM theme research argues that: (a) the metering of farm power supply to 14 million electric tube wells-the solution most widely espoused-poses a formidable logistical challenge and faces strident, mass-based farmer opposition which would make it politically difficult to implement quickly; (b) even if it is accepted, the logistical problems and high transaction costs of metering and billing a large number of dispersed farm power connections-which obliged governments to shift from metered tariff to flat tariff during the 1970s in the first place-remain on a far larger scale today; (c) if metering is to be introduced, its chances of working depend critically on privatization of metering and billing at the feeder level or below as has happened in China; and (d) however, in the short run, the best course is to transform the existing *degenerate* system of flat tariff into a *rational* flat tariff. This involves two things: first, raise flat tariffs moderately and regularly rather than in big jumps; and second, implement a proactive power supply policy for the farm sector by capping total hours of power supply over the entire year to a viable level relative to the level of flat tariff, but then schedule power supply to fit farmers' irrigation needs as best as possible. Pursuing this strategy of proactive management of rationed power supply can reduce power industry losses from its farm operations, reduce overall technical and commercial losses of power, curtail wasteful use of 12-20 km^3 of groundwater/year, and actually improve farmer satisfaction with the power industry (Shah, Scott *et al.* 2004).

5.5 CONCLUSIONS

Knowledge development in groundwater use in agriculture has so far remained asymmetric: while the world has learnt much about the hydrology of groundwater, it has learnt little about its social and institutional dimensions. Fifty years ago, when the challenge was to use groundwater to produce food and livelihoods, this asymmetry did not matter. Today, however, agricultural use of groundwater in some of the most populous regions of the world is rapidly surpassing limits of sustainability. In this new era, understanding the social science of groundwater irrigation is becoming as critical as understanding the resource itself.

Over the past 5 years, IWMI's Sustainable Groundwater Management theme has tried to fill this critical knowledge gap. In particular, it has contributed new research on three key aspects: (a) evolving the 'big picture' of groundwater irrigation; (b) analyzing technologies, institutions and local management approaches that enhance net social benefits of groundwater irrigation; and (c) exploring a practical toolkit of 'sustainability solutions'. In working on such a large mandate, the SGM theme has focused on regions of Asia, namely South Asia and North China, where the scale of groundwater use in agriculture is unparalleled elsewhere in the world and where resource governance issues are rapidly coming to a head.

The SGM theme has applied a rather broad and 'inclusive' notion of 'policy research' to better understand the groundwater challenges facing Asian governments, and has tried to fuse resource, user and institutional perspectives towards a well-rounded understanding of the groundwater socio-ecology in the region and its implications for other parts of the developing world. Its research on the 'big picture' of the groundwater economy has been an important building block in this regard. For example, the Indian discussion on water use in agriculture has been dominated by physical modeling and location-bound work, and oblivious to socioeconomic, institutional and policy dimensions of the burgeoning national groundwater economy. SGM research has focused on this economy, its impact on livelihoods and farm productivity, working of groundwater markets and other institutions, and explored totally new hypotheses about how this crucial economy rises and falls over time, and with what socio-ecological fall out. Such work can have important impact in countries like Sri Lanka, Vietnam and even some African countries where the size and significance of the groundwater economy are seriously underestimated.

The study of institutional interventions is another area where the SGM theme has contributed by addressing such questions as: What might be effective approaches for rebuilding communities' stake in improved management and upkeep of hundreds of thousands of tanks that were once the mainstay of their livelihood systems and now of their groundwater-based irrigation economy? How best to improve the performance of public tube well programs? And, how important is to engage local communities in managing their water resources sustainably? Together with this, IWMI has also examined various ways for governments to manage groundwater use in agriculture. Through this research, IWMI has highlighted and promoted strategic, indirect means for managing groundwater economies, such as through the energy-irrigation nexus, which in the short run may hold more promise in many developing countries than direct regulation.

As a result of these endeavors, the SGM theme has made significant progress towards its goal of providing a more precise understanding of the socio-ecological issues surrounding groundwater and aggressively promoting promising solutions for its sustainable use in developing countries. While IWMI's research on groundwater management is far from complete, having a specific theme devoted to the issue, has produced the intended result of highlighting the distinct and significant role of groundwater in agricultural production and rural livelihoods as well as the potential socioeconomic and environmental consequences of its unsustainable use. With this strong foundation, IWMI's new research framework no longer separates research on surface water and groundwater management but rather integrates the two in an effort to promote a more holistic approach to agricultural water management. As described in more detail in the concluding chapter, IWMI's groundwater research has been incorporated into the Institute's four new themes, which together assess issues of water productivity, water poverty and high potential interventions to manage agricultural water resources, across the hydrologic cycle, more productively, equitably and sustainably.

6

WATER RESOURCES INSTITUTIONS AND POLICY

Madar Samad

6.1 INTRODUCTION

It is widely recognized that major problems in the water sector are mainly caused by failures in its governance system relating to water laws, regulations and institutions (see Cosgrove and Rijsberman 2000; World Bank 2003). This has prompted countries to search for new and more effective policies and institutional arrangements that will try to balance the interests of all stakeholders while ensuring food security, alleviating poverty and protecting vital eco-systems. Now, there is also an international consensus on the subject as reflected in terms of a number of international declarations and agreements.[1] One significant feature is that all international covenants manifestly identify policy and institutional reforms for effective water governance as one of the highest

[1] Agenda 21, World Water Forum 1-3, the Millennium Development Goals adopted by the United Nations Millennium Summit in September 2000, Bonn Freshwater Conference in 2001and the Johannesburg Summit on Sustainable Development in August 2002 are some of the more prominent international events.

ISBN: 9781843391128. Published by IWA Publishing, London, UK.

priorities for action. Added to this is also a clear recognition that neither the state nor the market can by itself resolve the water management challenges. As a result, there are no blueprints for a universally applicable framework for water governance but it has to be tailored to suit local conditions with the benefit of lessons from international experience (Rogers and Hall 2003). The urgent challenge now is to explore the whole subject of water governance so as to translate it into practically applicable principles.

Research on institutions and policies has occupied a prominent place in IWMI's research agenda since the inception of the organization. In the early days, the focus was primarily on three sets of activities: (a) improving the performance of public organizations that manage irrigation systems through improved designs, operational procedures and performance assessment methodologies; (b) understanding the institutional arrangements and management practices of the enduring indigenous farmer-managed irrigation systems; and (c) analyzing the external and internal stresses that constrain current performance of these organizations. During its first decade, IWMI (then known as IIMI) did extensive work in more than 15 countries on these topics (Vermillion *et al.* 1996; Merrey 1997). But, in 1999, the Policy, Institutions and Management program was established to broaden the focus on institutional and socioeconomic research at IWMI. Under this program, policy options for optimizing water productivity and issues relating to water institutions and organizations, poverty alleviation and food security, gender analysis, and inter-sectoral conflicts have received more focused treatment.

As a logical extension and succession to this program, the theme on Water Resource Institutions and Policy (WRIP) was established in 2001. Its idea was to further sharpen the focus on issues directly relating to water policies and institutions and undertake studies to facilitate the design, adoption and implementation of effective governance frameworks for integrated water resources management at subbasin scales. Through systematic comparative research and capacity building activities, the WRIP theme aims to produce knowledge-based guidelines and best practice cases in policies, governance frameworks and organizational designs that can improve water and land productivity, enhance food and livelihood security and sustain the environment. The research under the WRIP theme has been organized under five broad areas: (a) institutional reforms in the water sector, (b) institutions for river-basin management, (c) water and poverty, (d) economic issues, and (e) gender and social issues. This chapter provides a summary and synthesis of IWMI research on water policies and institutions under these five areas against the backdrop of a quick review of earlier institutional research focused largely on irrigation management reforms.

6.2 IRRIGATION MANAGEMENT TRANSFER: REQUISITES AND IMPACTS

As discussed in Chapter 2, IWMI's research mandate was broadened from 'irrigation management' to 'water management' in 1996. As a result, the research canvas was changed from the irrigation systems to river basins with an obvious change in the agenda of institutional and policy research at IWMI. However, IWMI had, and continues to have, a major impact on global water research and policy in terms of its concerted works in the area of irrigation management reforms. This section reviews these works that constitute the early and the formative stage of institutional research at IWMI. These works covered a wide range of topics, including farmer participation in irrigation management and irrigation management transfer (IMT). The research at this stage was guided by some of the influential works on the subject (e.g., Wade 1982; Uphoff 1986; Chambers 1988; Ostrom 1989). During this period, IWMI research on the subject was also driven by a belief that IMT could empower farmers and users. Studies on irrigation management reforms dominated IWMI's research agenda in the 1980s and early 1990s and the volume of work done in this field is substantial. The main focus of this research was on the preconditions for viable transfer programs and their impacts on the performance of irrigation system and irrigated agriculture.

6.2.1 IMT: Preconditions and viability

A successful IMT program requires fulfillment of a well-defined set of tasks and the installation of complementary institutional elements so that management transfer will result in effective services to water users at a reasonable cost (Frederiksen and Visia 1998). A number of IWMI works (e.g., Vermillion 1997, 1998; Brewer *et al.* 1997; Johnson 1997) have tried to specify the preconditions for the successful transfer of management of irrigation services based on empirical case studies from different countries. On the basis of comprehensive reviews of research, Vermillion (1997) identified the following prerequisites:

(a) clearly recognized and sustainable water right and water service
(b) compatibility of infrastructure with water service and local management capacities
(c) well-specified management functions and assignment of authority
(d) effective accountability and incentives for management
(e) viable systems for timely conflict resolution
(f) adequate resource mobilization for irrigation management

Based on field studies in the Tambraparani Irrigation System in Tamil Nadu, India, Brewer *et al.* (1997) suggest three basic provisions are needed for successful transfer: recognition of water user associations and joint management committees; transfer of legal authority to these associations; and formal legalization of the changes in water allocation procedures. Merrey (1997) addressed the question of how to design institutions accountable for managing large-scale irrigation schemes. Based on a review of selected case studies and drawing largely on the earlier works of Coward (1986) and Ostrom (1992, 1993), he argues that where each irrigation scheme is managed by an autonomous organization accountable to its customers, performance will be better than those managed by agencies dependent on the government or by agencies managing large numbers of systems as in Asia.

6.2.2 IMT: Impacts and assessments

Considering the lack of studies on the impacts of IMT, especially from a comparative international perspective, IWMI, in 1996, initiated a series of case studies to assess the impacts of IMT. The impacts were evaluated considering aspects such as equity, operational efficiency, cost recovery, agricultural productivity and sustainability of irrigation systems. The evaluation involved a comparative review and analysis of the findings from 29 research studies of IMT (Vermillion 1997). This was also followed by a series of country-specific studies in Asia, Latin America and, to a limited extent, in Africa (e.g., Vermillion and Johnson 1995; Samad *et al.* 1995; Kloezen *et al.* 1997; Johnson 1997; Vermillion *et al.* 2000; Abernethy *et al.* 2000). While the earlier studies varied in conceptual design, those carried out in later years adopted a common methodology to facilitate comparative analyses (see Vermillion *et al.* 1996). Most studies relied on farm/system level data and used either 'before-after' or 'with-without' kind of approaches in evaluating the impacts of IMT. Approaches based on stakeholder and *post-facto* assessments are also used in the evaluation of single cases.

The results suggest that with the exception of Mexico and Colombia, there is insufficient evidence that IMT has had any positive impacts. In most Asian countries, the main change was a gradual decline in government funds for the O&M. With insufficient users' contributions for maintenance, there are concerns as to the long-term sustainability of irrigation systems. Again, with a few exceptions, there is no discernible evidence for the impacts of IMT on system operations and agricultural production. Evidence relating to agricultural productivity is, however, mixed. In the Lagunera region in Mexico, an assessment of performance of two transferred modules revealed that water users were successful in implementing crop plans and water allocation without any head-tail or other biases (Levine *et al.* 1998). In Colombia, IMT had improved management efficiency and agency accountability but it had only a neutral effect on agricultural productivity (Vermillion and Garcés-Restrepo 1998). Despite having no, or limited, impacts on agricultural or water management performance, IMT did contribute to improved communications between farmers and officials, increased responsiveness of staff and reduced the hassles of arranging water deliveries and paying of water charges.

In Africa, IMT works covered the experience in Niger, Nigeria and Sudan. In Niger, the attention was on the institutional and financial viability of systems transferred to farmer cooperatives (Abernethy *et al.* 2000). In Nigeria, IWMI worked with a river-basin management authority to pilot-test a participatory action research approach for organizing farmers based on its experiences in Asia. Although the initial results were promising, they were not conclusive due to short project periods (Merrey 1997). In Sudan, the focus was on the viability of the transfer of pump irrigation schemes to farmers along the White Nile (Samad *et al.* 1995). Unfortunately, this experiment was reduced to a mere transfer of schemes from a public monopoly to a private oligarchy due to a flawed policy implementation.[2] Shah *et al.* (2002) have addressed the larger question of whether it is possible to replicate the successful IMT experiences in other continents in the African context. The replication will not be that straightforward, especially given the predominance of smallholder irrigation in Africa. IMT is unlikely to work in African smallholder irrigation, unless the trajectory of productivity and income is raised substantially in the first place. It is only with such a higher income that the smallholders will be able to bear the additional costs and management responsibilities. Therefore, IMT has to be preceded by institutional changes needed to relax the complex set of economic and technical constraints of smallholder systems.

6.3 INSTITUTIONAL REFORMS IN WATER SECTOR

In recent years, IWMI has also generated a few but important works on the analytics and dynamics of institutional reforms within the water sector, especially from an institutional and political economy perspective. These works cover both the development of analytical and methodological frameworks as

[2]As the state was more preoccupied with the divestiture of the parastatal agency than with the perusal of complementary policies and supportive institutional reforms, the transfer program was devoid of the required incentives. Despite the transfer program, institutions continued to be oriented to a state-dominated economy with high agricultural taxes, heavy regulation of the cultivation of the main irrigated crops (cotton and wheat), and poorly developed private markets for farm inputs and services (see Samad *et al.* 1955).

well as the practical application of these frameworks for understanding the dynamics of factors that influence reforms in different country and sectoral settings.

6.3.1 Methodologies for evaluating institutional reforms

For the purpose of understanding institutional reforms, water institutions are defined as an interactive structure determined by the prevailing water law, water policy and water organizations. This internal structure of water institutions is, then, distinguished from their external environment, which is characterized both by the physical setting of the water sector and by the social, economic and political setting of the country (see Saleth and Dinar 2004). This way of characterizing water institutions allows one to develop a framework for a two-stage institutional decomposition. First, the external and internal aspects of water institutions are distinguished. That is, the water institutional structure (governance structure), as determined by water-related law, policy and organizational elements, is delineated from the water institutional environment (governance framework) that is determined by the historical, constitutional, economic, social, political and physical conditions of the country. Second, the water institutional structure is decomposed as water law, water policy and water organization and each of these institutional components is, in turn, further unbundled to highlight a few of the most important law, policy and organization-related institutional aspects or rules. Such a framework of institutional unbundling has major analytical and theoretical advantages. It permits to visualize the internal linkages and dynamics of water institutions and also enables to trace the influences of both the endogenous and exogenous factors within an institutional transaction cost framework.

Saleth (2004) has used the same approach to develop an institutional transaction framework and applied it to explain the process of water institutional reforms in India. The nature, extent and coverage of institutional reforms in India clearly provide evidence for the powerful effects that exogenous factors (e.g., economic liberalization policies, political forces, international financial and research institutions, and natural calamities) have on the opportunity and transaction costs of institutional change within the water sector. Notably, the initiatives undertaken initially involved only the transaction cost-wise easier and ceremonial options (e.g., declaration of a water policy, constituting committees and marginal legal amendments). However, those undertaken in recent years involved politically difficult and substantive options (e.g., administrative reforms, basin organizations, irrigation management transfer, and the promotion of autonomous corporations and private-sector involvement). But, India is yet to move to the stage of embarking on real reforms (e.g., review of the center-state relation in the water sector, declaration of an exclusive water law, creation of water rights system at various levels, and administrative reforms for water sub-sectoral coordination, staff resizing and balanced functional specialization). Understandably, these reform options involve heavy economic and political transactions costs. Although these costs are lower than the potential performance benefits, the differential weights assigned by political leaders often distort the transaction cost calculus.

6.3.2 Cross-country analysis of institutional reforms

Saleth and Dinar (2004) applied the analytical framework based on the institutional unbundling approach for a quantitative analysis of water institutions and their performance. They utilized data collected from 117 water experts from 36 countries around the world. The results provide evidence for upstream and downstream linkages across institutional components and indicate how these linkages can be strategically used to counter political and technical constraints for reforms.[3] This means that the linkages among institutional components and the synergies from exogenous factors provide a basis for developing reform design and implementation principles, such as institutional sequencing and packaging, reform timing and spacing, and program scale and coverage. To investigate to what extent these principles are used in actual reforms and identify the relative role of the practices of endogenous and exogenous factors in prompting reforms in country contexts, IWMI has organized a set of six papers each of which addresses the same questions in six different countries: Australia, Chile, Morocco, Namibia, South Africa and Sri Lanka.[4] Saleth and Dinar (2005) have provided a synthesis of the main findings from the reform experiences of these countries with the hypotheses of institutional reform theories, especially those from the political economy and institutional transaction cost theories. Some of the results are presented in Tables 6.1 and 6.2.

As to the configuration and relative importance of factors behind the reform process, Table 6.1 presents the results in terms of some of the major factors for each of the six sample countries. These factors are identified with a diagnostic

---

[3] For instance, the downstream linkages among user organization, water rights and water markets ensure that the development of water markets can be facilitated by the prior creation of user organizations and water rights. Thus, from a reform perspective, there is no need to create these institutional components all at the same time. They can be created in sequence, taking the politically easier components first and exploiting the institutional synergies better through appropriate timing and spacing of different components. For more details, see Saleth and Dinar 2004.

[4] These papers, along with the introduction and methodological papers, are presented in a special issue of Water Policy 2005, 7(1).

use of the institutional transaction cost framework, i.e., by considering the probable effects of different factors on the transaction and opportunity costs of change. As can be seen, in almost all cases, endogenous factors (water scarcity, sectoral financial crisis and droughts) have remained the fundamental force for change, though it is the exogenous factors (economic and political reforms) that provided the immediate prompt for reforms. On the other hand, the scale economy-related internal institutional synergies and pressures are important only at a mature stage of institutional reforms.

Table 6.1 Configuration and role of factors behind water institutional reforms.

Factors	Australia	Chile	Morocco	Namibia	South Africa	Sri Lanka
Water scarcity/ conflicts	**	*	**	**	**	*
Financial crisis	*	**	**	***	*	***
Draughts/ salinity	***	-	***	*	**	-
Macro-economic reforms	***	**	***	-	-	***
Political reforms	-	***	-	***	***	*
Social issues	*	-	*	**	**	-
Donor pressures	-	*	**	*	-	***
Internal/ External agreements	***	-	-	*	*	-
Institutional synergy/ pressures	**	***	*	*	*	*

Note: The number of *s signifies the perceived relative importance of the factors in the context of each country. '-' means the aspect in question is 'not applicable' or 'not evaluated.'

Source: Saleth and Dinar 2005.

Since the institutional change process is not entirely evolutionary or autonomous, there is a vast scope for deliberate and purposive policies, including the use of reform design and implementation principles. The reform experiences of the six sample countries provide considerable evidence not only for the actual adoption of these principles but also for the relevance of some of the theoretical postulates on the influence of stakeholder perception and the existence of political bargaining (see Table 6.2). As can be seen, most of the countries have used the reform principles, though with a differential emphasis. The reform intensity is relatively high in South Africa and Australia; the focus on the latter is on higher-level reforms in contrast to the initial stage reforms in the former. Stakeholder perception has played a significant role in the reform process in Australia, South Africa and Sri Lanka though there is an obvious difference in its effectiveness to create the necessary pressures for reforms. The same fact also applies to political bargaining, as there has been a successful outcome in two cases (Australia and South Africa), but only a deadlock in the other case (Sri Lanka).

Table 6.2 Relevance of theoretical postulates and reliance on reform principles.

Particulars	Australia	Chile	Morocco	Namibia	South Africa	Sri Lanka
Intensity of reforms	4	3	3	2	4	1
Perception/ pressures	2	-	-	-	3	3
Political bargaining	3	-	-	-	2	1
Reform packaging	3	-	2	1	3	1
Reform sequencing	3	2	2	-	2	-
Reform timing	4	-	2	2	4	-
Reform scale/ coverage	4	3	3	2	4	1
Scope for scale economies	4	3	2	-	3	-

Note: All numbers indicate the perceived level of significance on a scale of 1-5. They indicate only the relative importance across countries. '-' means the aspect in question is "not applicable" or "not evaluated".

Source: Saleth and Dinar 2005.

Although the demand-side role of institutional education and the supply-side role of comparative research are now being increasingly recognized, they are not yet getting the policy attention they deserve. This is because of the mistaken view that their effects are slow, remote and marginal. But, considering the fact that ambiguity in the understanding and divergence in the interpretations of institutions often constitutes the initial but main stumbling blocks for institutional reforms in many contexts, the institutional roles of education and research can be immediate, substantial and indispensable. The way water institutions and their change process are conceptualized can be a starting point for developing institutional learning and evaluation tools to facilitate a better and consensual understanding of institutions among both the public and policymakers.

6.4 INSTITUTIONS FOR RIVER-BASIN MANAGEMENT

With evolution of the overall research focus from irrigation management at system level to water management at the river-basin context, issues such as organizational options, rules and support systems for effective river-basin management have emerged as priority areas for research at IWMI. The first major study to address these issues was carried out under the two interrelated projects, *Institutional Support Systems for Sustainable Local Management of Irrigation in Water-Short Basins* (ISSP) funded by BMZ through GTZ of the Federal Republic of Germany and *Developing Effective Management Institutions* supported by the Asian Development Bank. The first project, launched in 1998 as a 3-year effort, initially considering three river basins, i.e., Lerma-Chapala in Mexico, Gediz in Turkey and Olifants in South Africa. The second project, launched in 1999, examined basin institutions and their interactions with local irrigation-management structures in six river basins in Asia (China, Indonesia, Nepal, Philippines, Sri Lanka and Thailand). It has also considered two additional case studies of river basins in Japan and Australia. The key research question addressed in both projects was: in water-scarce basins with a significant locally managed irrigation, what tasks must be carried out at the basin level and what are the most appropriate techniques and institutional arrangements for carrying out such a task?

6.4.1 Hydro-institutional mapping

The study under ISSP has tried to apply and validate the new methodologies and tools that IWMI had developed such as 'water accounting' and 'hydro-institutional mapping' (see Molden and Sakthivadivel 1999; Molden *et al.* 2005). While water accounting is easy to understand, hydro-institutional mapping can be explained as an attempt to chart the development phases of a river basin and then, identify the essential institutional requirements in each phase. An effective institutional arrangement is one that has the ability to cater to the evolving management needs of the basins as water use changes over time. The basin-level institutional mapping exercise can be explained using Figure 2.1 (see Chapter 2) that depicts the three development phases of a river basin, i.e., development, utilization and allocation. The key features, concerns and priorities of these phases can be summarized as in Table 6.3.

Table 6.3 Phases of river basin development: Key features and main concerns.

Characteristic	Basin-Development Phases		
	Development	Utilization	Allocation
Dominant activity	Construction	Managing supplies	Managing demand
Utilizable flow depleted (Fraction)	Low (0 to 0.4)	Medium (0.4 to 0.7)	High (0.4 to >1.0)
Value of water	Low value of water	Increasing value of water	High value of water
Infrastructure	Installing new structures	Modernization/rehabilitation	Measurement, regulating
Groundwater	Utilizing groundwater	Conjunctive management	Regulating groundwater
Pollution	Diluting pollution	Emerging pollution/salinity	Cleaning up pollution
Conflicts	Fewer water conflicts	Within system conflicts	Cross-sectoral conflicts
Water scarcity	Economic water scarcity	Institutional water scarcity	Physical water scarcity
Data needs	Perceived as less important	System water delivery data	Basin water accounting data
Water-poverty concerns	Including/excluding poor in development of facilities	Including poor in O&M decision making	Loss of access to water by the poor

Source: Molden *et al.* 2005.

As can be seen from Table 6.3, the institutional concerns differ depending on the phase of development. These concerns do exist at all times but their importance or emphasis may change over time. Institutions have to adapt to meet these changing concerns, as their functions needed at the management and allocation phases are entirely different from those needed at the development phase. But, it is widely agreed that institutions have to perform a set of essential functions in a reasonable manner for effective and sustainable river-basin management. Table 6.4 lists these functions. At the same time, good basin governance, effective resource mobilization and regular performance assessment are also essential to ensure these essential functions of institutions.

Table 6.4 River-basin management: Major institutional functions.

Functions	Details
1. Plan	The formulation of medium- to long-term plans for the management and development of water resources in the basin, by which the water demands of different sectors are brought in line with water supply.
2. Allocate water	The mechanisms and criteria by which water is allocated to the different use sectors, and by whom.
3. Distribute water	The activities executed to ensure that allocated water reaches its point of use.
4. Monitor water quality	The activities executed to monitor water pollution and salinity levels.
5. Enforce water quality	The activities executed to ensure that water pollution and salinity levels remain below accepted standards.
6. Protect against floods and other catastrophes	Activities executed concerning flood warning, prevention of floods and emergency works, as well as drought management, infrastructural failure, etc.
7. Protect ecology	Priorities and actions undertaken by the stakeholders in the basin to protect aquatic ecosystems.
8. Construct facilities	Activities executed for the design and construction of hydraulic infrastructure.
9. Maintain facilities	Activities executed to maintain the hydraulic infrastructure in the basin.

Source: IWMI 2001b.

6.4.2 Cross-basin comparative analysis

The Asian river-basin study has initially considered five basins: Fuyang River Basin in the People's Republic of China, Singkark-Ombilin in Indonesia, Upper Pampanga in the Philippines, East Rapti River Basin in Nepal and the Deduru Oya in Sri Lanka. Since the five sites chosen reflect a full range of the basin development phases, they allow a cross-site comparison to develop a perspective on the issues occurring in various phases of basin development (see Sakthivadivel and Molden 2002). In 2001, two additional basins from Thailand, Mae Klong and Bang Pakong, were included. Besides these seven cases, three more cases, considered to be 'success stories' were also included as supplementary cases. Two of them, Murray-Darling in Australia and Omonogawa in Japan, were considered to add contrast to, and derive lessons from, the experiences in developed countries. The third case study was from the Brantas Basin in Indonesia. The focus of the cross-basin comparative analysis was on the question of water resources availability for agriculture in the context of growing inter-sectoral competition and on the associated environmental, socioeconomic and institutional issues. The details of the analytical framework used in this comparative study are given in Bandaragoda 2000

The institutional conditions for stakeholder participation have received particular attention in all cross-basin comparative studies conducted by IWMI. However, in the study by Wester *et al.* (2003), this issue received an exclusive focus, especially from a political economy perspective. By comparing the process of stakeholder representation in two river basins, the Lerma-Chapala in Mexico and the Olifants in South Africa, this study evaluated the basic premise that since water is a politically contested resource, water management institutions and policies are heavily influenced by political practices. Under this condition, it is important to analyze how power pervades through institutional arrangements causing differential access and control over water and what mechanisms are needed to redress such inequities.

The study shows that river-basin management in Mexico is a top-down and state-driven process with attempts being made to involve stakeholders in the decision-making process. In South Africa, the thrust of the new legal and policy changes was to replace the past system of centralized management by the Department of Water Affairs and Forestry (DWAF) with decentralized management at the basin level through the creation of the Catchment Management Agencies (CMAs). The Mexican experience is relevant in conditions where the major stakeholders are well organized, as is partly the case in Mexico, or where economic growth provides alternative livelihood options for the poor. The South African experience, in contrast, is more relevant to developing countries that are considering new policies and institutional arrangements for river-basin management.

6.4.3 Key results from basin institutional analysis

While more details on the two cross-basin comparative studies can be found in Svendsen 2005, Bruns *et al.* 2002, and Bruns and Bandaragoda 2003, here some key results are provided as stylized facts.

(a) There are clear stages in the development of river basins, which can be characterized as development, management and allocation. These stages can be identified in terms of the fraction of the total available water committed to various uses at a given period.

(b) The rive-basin development framework is a useful tool for identifying gaps in the management structure of water-scarce river basins and assessing the effectiveness of the existing institutional framework in relation to the institutional functions needed at different stages.

(c) One key institutional component that is critical in water-short basins is the creation and enforcement of a system of clearly defined water rights both to protect the poor and disadvantaged as well as to ensure adequate environmental flows to meet sustainability goals. Unfortunately, this is missing or poorly developed in most river basins in developing countries.

(d) The cross-basin comparative analysis of the cases from developing and developed countries suggests that formal 'river basin organizations' need not necessarily be an essential feature of the successfully managed water-scarce river basins. Other arrangements, including the legal systems and various kinds of committees and networks, can often work just as effectively.

(e) There is a clear need to create effective mechanisms for stakeholder consultations and enlist their cooperation in implementing programs for developing and managing water resources. The experiences from basins suggest that well-designed and stakeholder-driven institutions are more likely to have positive outcomes.

(f) The three 'success stories' of basin management in Murray-Darling in Australia, Omonogawa in Japan and Brantas in Indonesia suggest that institutional development has been a slow process, taking decades to become functional and effective. This suggests the need for more research on the emergence of appropriate institutional arrangements and the sequence in which new arrangements should be introduced.

(g) One important service or function within the river-basin context is drainage management, including the issues of water reuse. While there are significant attempts in developing practical methodologies and implementation strategies for drainage issues in the general water-sector context (e.g., Abdel-dayem *et al.* 2004; World Bank 2004), there are hardly any studies that integrate the institutional dimension of this major issue within the river-basin context. Clearly, there is a need for more research on this dimension of river-basin management.

(h) Finally, one factor that distinguishes successful cases of river-basin management from the rest is the presence of sound technical and hydrological information at basin and subbasin scales. What is also important is the extent to which such information is shared among stakeholders in the basin and the way they guide water allocation and use decisions. The institutional arrangements for data generation and their transformation into key management inputs remain a priority issue in the context of river-basin management, especially in water-short basins in developing countries.

6.5 WATER AND POVERTY

Water has a varied and wide-ranging impact on poverty alleviation in terms of its direct and indirect impacts on the generation of rural employment, income and livelihoods. Since irrigation plays a more dominant role in this respect, the research focus of IWMI was largely on the linkages between irrigation and poverty. But this focus was broadened over the past few years to cover also the poverty and livelihood effects of nonirrigation uses, including environmental flows for supporting riverine and wetland ecosystems as well as multiple uses and water-reuse practices.

6.5.1 Irrigation and poverty

The poverty alleviation role of irrigation development has been the central focus of a major project undertaken by IWMI with funding support from the Asian Development Bank and the Japan Bank for International Cooperation. This project involves a series of in-depth case studies conducted in six Asian countries: Bangladesh, China, India, Indonesia, Pakistan and Vietnam. The main message emanating from these case studies is that irrigation development is a major contributory factor in reducing poverty but, that its anti-poverty impacts are greatest in settings where there are less socioeconomic differentiations and more equity, especially in terms of land tenure and water allocation. Some of the key results of this multi-country study (Hussain 2005) are presented below.

(a) Poverty outside of irrigation systems is much higher (almost twice) than within irrigation systems. Poverty levels are the highest in marginal areas, downstream sites and areas where canal water is in short supply and groundwater is low or of poor quality.
(b) The extent of benefits to the poor depends on factors such as land and water distribution, irrigation quality, input and infrastructural support and water and agricultural policies.
(c) Inequity and insecurity in access and rights to land and water are bad for both productivity and poverty. Where land and water equity exists, irrigation in itself is pro-poor (as in China and Vietnam).
(d) There are strong linkages between irrigation, gender diversity and poverty issues. In South Asian systems, poverty is generally higher among female-headed and low-caste/ethnic minority households.
(e) Irrigation systems managed by public agencies tend to perform poorly. Lack of clear and secure water rights and allocation rules and corruption-related problems adversely affect performance of irrigation systems and their poverty-reducing impacts.

(f) Across countries, poverty is higher in South Asian systems (particularly in Pakistani systems) than in Southeast Asian and Chinese systems, with inter-system differences in poverty much higher in the former than in the latter. Overall, South Asia has only partially benefited, in terms of realizing poverty-reducing impacts of past irrigation investments and hence, significant opportunities are still there for increasing benefits of irrigation.

(g) In South Asia, institutional reforms in irrigation sector are very slow and only on a limited scale (e.g., mostly at the tertiary 'canal' level but not much at higher levels). Unless irrigation reforms are sharpened with a pro-poor focus, the poor are likely to be bypassed.

(h) Irrigation investments have typically centered on the creation of physical facilities and institutions and on their economic performance in terms of aggregate costs and benefits, with little or no attention to specific benefits and costs to the poor.

(i) As a whole, South Asia has much to learn from experiences in land and water distribution, institutional management and technological interventions, in Southeast and East Asia, particularly China. In these latter regions, irrigation management and other support services are more incentive-based and relatively more equitable, and the agricultural productivity and the benefits of irrigation are high.

Besides the major study of Hussain (2005), there are also a few studies addressing the same in more specific contexts (groundwater and wastewater use), particularly using time series and cross-sectional data from India. For instance, using a panel data set for 14 Indian states over 1970-94, Bhattarai and Narayanamoorthy (2003) have attempted an econometric analysis of the relationship that poverty has with irrigation and other factors such as fertilizer use, high-yielding varieties, education and rural road density. The results support the fact that irrigation, especially from groundwater, is a major factor contributing to poverty alleviation.

Another study by Saleth, Namara and Samad (2003) has assessed the impact of irrigation on poverty, using a set of simultaneous equations and data related to 80 agro-climatic subzones of India for two time points, 1984-85 and 1994-95. While their results support a positive impact of irrigation on poverty alleviation, they also suggest a declining and changing nature of such impacts. Specifically, the impact of irrigation, which was initially in terms of labor absorption from area expansion and cropping intensity, tends to be now more in terms of increasing labor productivity and income, and falling food prices. Shah and Singh (2003) arrive at a similar conclusion based on their analysis of the data for 177 predominantly rural subdistricts of Gujarat collected from the Census of the

Below Poverty Line and the Village Amenity Survey conducted by the state in 1997. Data showed that subdistricts with higher irrigated area had a higher proportion of households below the poverty line. While the major route by which irrigation affects income and poverty is through intensified land use, such beneficial impacts at the farm or command level mask the fact the intensive irrigation development also acts as a magnet attracting poor people from surrounding areas ultimately resulting in more poor people in the area.[5]

From the perspective of water deprivation, wastewater plays an important role. The poverty alleviation roles of waste and reused water have also been addressed in a few recent studies in the context of the Musi River in Andhra Pradesh, India (e.g., Buechler and Devi 2003). Research has found that along this river that flows year-round with wastewater in semiarid regions in the outskirts of the Hyderabad City, wastewater use is extensively used not only for irrigating paddy, vegetables and fodder crops but also for supporting agroforestry and aquaculture. These wastewater-based livelihoods depend on caste and gender-related factors. What is interesting is the innovative nature of the coping practices that the wastewater users adopt, such as the crop pattern changes and mixing of groundwater with wastewater before irrigation.

6.5.2 Water deprivation and water entitlements

Overall, the results of IWMI studies and others in existing literature (see Saleth *et al.* 2003) suggest that access to irrigation water has tremendous potential to improve the livelihoods of the poor. Historically speaking, irrigation has played an important role in poverty alleviation. It has strengthened food security thanks to its positive role both on the supply side (increased output) and on the demand side (reduced food prices). It has also increased rural livelihoods with expanded opportunities for on-farm and off-farm employments. Development of large-scale canal irrigation has been an engine of regional development and economic growth in most Asian countries. But, the poverty reduction impacts of irrigation are not a foregone conclusion, as the irrigation landscape is changing.

With a growing scarcity and competition for water and conflicts among its uses and users, the poverty alleviation role of water is now under severe threat (Barker *et al.* 2000). The most accessible and cheapest water resources have been developed and there is now an intensive pressure for moving water away

---

[5] On surface, this appears to be counterintuitive as it means irrigation adds more poor people while reducing the same in other areas. A partial explanation lies in the idea of agricultural involution (see Geertz 1963) that allows an increasing absorption of labor force into a, more or less, static socio-technological structure, though at a mere subsistence level. With the labor-intensive production methods, this process may increase land productivity but with the same or low labor productivity.

from irrigation into the urban and environmental sectors. The generous subsidies provided to irrigation and agriculture are no longer sustainable. The prices of cereals, the most dominant crop grown under irrigation, are declining and increasing the pressure for crop diversification and new irrigation technologies (see Pinstrup-Andersen and Rajul Pandya-Lorch 2001). Under this condition, the poverty alleviation impacts of irrigation in particular and water in general are not automatic but require considerable support from complementary policies and institutional reforms.

Water entitlements are critical for eliminating water deprivation.[6] It is no wonder that most rural poor are now viewing the entitlement and access to water for productive and domestic uses as much more critical than the same for health care and education (Barker *et al.* 2000). Thus, the next stage of research on the subject should move from the focus on the direct first-round effects of irrigation towards the fundamental effects of water on the asset and resource base of the poor. It is against this backdrop that IWMI is adopting a comprehensive strategy for a holistic management of water, which is epitomized by the approach, 'from bucket to basin'. This approach, based on a bottom-up process, is articulated now as a principle for managing river basins to alleviate water deprivation, assure both efficiency and equity in water use, protect the interests of the poor and marginalized economic and social groups, and ensure the environmental sustainability of basin water systems (van Koppen 2000). In this approach, the focus is not only on the poverty and livelihood roles of irrigation but also on similar roles of water allocated to wetland and other water-based ecosystems.

6.6 ECONOMIC ISSUES

Over the years, IWMI has generated a large body of research, addressing some specific but major economic issues relating to irrigation in particular and water in general. It covers not only such issues as water pricing, water productivity, irrigation investment and cost-benefit analysis but also the nature and implications of global water demand-supply scenarios.

6.6.1 Water pricing

Water being a finite resource with increasing and conflicting claimants, there is considerable justification for treating it as an 'economic good' as proposed

[6] "Water deprivation" is viewed primarily as human-made and not as the inevitable result of natural scarcity. It is conceptualized as both "asset-related" (technological, institutional, and financial limitations of the society to deliver water) and "direct deprivation" (rich disenfranchising the poor). For details, see van Koppen 2000.

indeed in the Dublin Principle. But, given its poverty alleviation and basic need roles, it has also to be treated as a 'social good'. Depending on the nature and context of its use as well as the technological and institutional conditions, water also shares the features of both a private (economic) and a public (social) good. These aspects are to be taken into consideration while addressing the role of water pricing in the context of its two economic roles: resource allocation and cost recovery. It is within this broad context that IWMI research on the issues is conceptualized (Perry *et al.* 1997). While recognizing the importance of the pricing instrument, its conceptual and practical challenges, especially in the context of the multiple roles and attributes of water, cannot be underestimated.

Since water market prices often reflect the financial, not necessarily the economic, values, they can create serious equity issues. Apart from the conceptual and equity issues, there are also other practical difficulties in applying the pricing instrument when the resource flows through complex structural and distribution networks within a basin, where return flow issues, externalities and free-riding can be rampant. Besides the potential for market failures, the transaction costs can also be too high to be socially acceptable. While appropriately designed and applied market instruments can be expected to yield benefits, the necessary and sufficient conditions for these markets, especially legally defined and locally enforceable water rights, are not yet in place for these anticipated benefits to materialize (see Perry *et al.* 1997; Rogers *et al.* 1997). Even though there are difficulties for the allocation role of water pricing, there is a clear scope for its cost recovery role, especially given the low water rates and poor cost recovery observed in most developing countries. Attention should be more on this potential than on the dogmatic debates over whether water is an economic good or a social good. From a practical perspective, water policy formulation should be done within a multi-objective decision-making framework with a clear recognition that the social and economic values of water will vary substantially across contexts and uses as well as over time and space.

As to the issue of cost recovery, IWMI research has initially focused on the question of whether farmers can afford to pay for irrigation O&M services. Results from Egypt and Sri Lanka indicated that since the O&M costs amount only 4.0 to 6.5% of the net farm incomes, farmers could afford to pay these costs. But, they would not be able to pay the full supply cost of irrigation. Going a step further, a modeling study carried out in Egypt has not only compared the relative effectiveness of three methods of water fee collection (flat area fee, crop-based charges and volumetric fees) but also examined the impact of volumetric fees on irrigation efficiency (Perry 1995). This study indicated that even if fees were raised to politically unrealistic high levels, they would not significantly reduce water use. Since the higher transaction costs of crop-based or volumetric charges would exceed the potential benefits, it is more reasonable for Egypt to use a flat area fee that is cheaper to implement and able to achieve cost recovery. Perry (2002) also reached the same conclusion in a recent study he conducted in Iran. Thus, the price level needed to influence irrigation demand in the study site in Iran is far beyond the politically acceptable range and even that will not be effective without a massive investment in physical, legal and administrative infrastructure needed for controlled water delivery. The study suggests an alternative approach of use-specific water rationing to cope with water shortages. This approach also has a number of potential benefits including simplicity, transparency and the potential to tailor allocations to hydrological situations, particularly where salinity is a problem (Perry 2002).

In a recent study, Dinar and Saleth (2005) have argued that by taking an institutional approach, most of the issues in the water pricing can be handled within a unified framework. That is, rather than the usual approach of viewing it as an economic, financial or equity instrument, water pricing has to be viewed as institutional configurations that determine the rules of payment and allocation. The discretionary or mark-up pricing is one of the institutional mechanisms that the society has created to deal with contexts where markets are absent due to high transaction costs or markets can exist but unable to address certain social and equity considerations. This is clearly the case with water, especially in irrigation and domestic uses and particularly so in the context of some social and economic groups. From an institutional perspective, the levels and methods of water pricing can be interpreted as forms of 'rules' to determine the payment and use of the resource in different contexts.

These rules do not exist in isolation but are structurally and operationally linked with other water-related legal, policy and organizational rules or mechanisms. In this sense, for performing its different roles, water pricing requires a different set of institutional conditions - with simple ones for cost recovery but the most complex arrangements for performing the allocation and equity roles. Besides the high institutional transaction costs for the allocation and equity roles, there is also an uncertainty as to whether a pricing system, which is efficient otherwise, can be consistent with the social, equity and environmental goals of water management. It is these institutional and political economy issues that are remaining as constraints for the water-pricing reforms.

6.6.2 The economics of water productivity

With increasing water scarcity, there is an incentive to adopt water-saving technologies and improve water productivity in the sense of 'more crop per drop'. The water productivity concept entails the notion of efficiency, though there is some confusion over the level and profit implications of such efficiency.

As a result, the level of efficiency in the surface irrigation context is often overstated. It is also important to recognize that water saving may not necessarily lead to gains in water productivity and that gains in water productivity may not lead to higher profits. Moreover, gains in water productivity at farm levels may not automatically translate themselves into gains at system or basin levels. As argued by Seckler *et al.* (2003) and Barker *et al.* (2003), the key to these problems lies in the way water productivity is defined. There are several ways of expressing productivity.

For instance, pure physical productivity is defined as the quantity of output divided by the quantity of available, diverted or depleted water, expressed as kg/m^3. Economic productivity, in contrast, is the gross or net present value of output divided by the net present value of the available, diverted or depleted water and it is defined in terms of its opportunity cost, the value of water in the next best alternative use. This concept of water productivity has been applied to address the effects of various policies and management practices on water saving and water productivity. For instance, Barker *et al.* (2001) have examined the relative effects of alternative water-saving practices on water productivity in the context of rice cultivation in China. In the same context, Hong *et al.* (2001) have evaluated the policies and management practices that have contributed to the gains in water productivity over time whereas Moya *et al.* (2001) have assessed the water savings and economic benefits due to the adoption of the practice of irrigating rice by *alternate wetting and drying*. Work is currently underway to assess the impact of a change in government taxation and fee collection policy on water saving and productivity.

6.6.3 Investments in irrigation development

Irrigation investments-both at the national and the global levels-witnessed a dramatic fall, especially compared to their levels during the 1970s and 1980s. For instance, the World Bank lending for irrigation, which averaged annually at around 7% of the total World Bank lending during 1960-90, has declined to just 2.5% of the total lending in the 1990s and has continued to decline further in the early 2000s as well (FAO 2003). There are a number of reasons for this decline such as the reduced potential for irrigation development in areas with heavy water depletion, competing claims from nonirrigation sectors, and the more stringent environmental rules and regulations. With the attainment of food self-sufficiency and low world prices for irrigated crops, there has also been a reduced political and economic pressure for irrigation expansion. While the poor performance of past irrigation projects has reduced the additional benefits, the cost of irrigation development has increased, as the most suitable sites tend to be exhausted. For instance, irrigation construction costs have risen to two to three times their previous level (FAO 2003). In short, the political economy of irrigation development is less favorable than in the past two decades. But, this does not mean that the full potential of irrigation development has been reached or the need to provide irrigation services to the poor farmers has been attained.

Kikuchi *et al.* (2002) analyze the trends in irrigation investment in Sri Lanka and a notable analytical feature of the study is the reliance on a stage-based typology of the agricultural development and irrigation investment. The study identifies three distinct phases in the history of irrigated agriculture in Sri Lanka.

In the first phase, land resources are abundant and agricultural output is increased by opening new land. The second phase represents the construction of new irrigation facilities as irrigation investment becomes more profitable with agricultural growth. The third phase involves enhancing the performance of existing irrigation systems through various interventions, such as rehabilitation, modernization and management reforms.

The nature and behavior of irrigation investment obviously vary during these three phases. Public investments on irrigation that commenced on a major scale in the mid-1970s had reached a peak in the mid-1980s, but had witnessed a sharp decline thereafter. But, in contrast to the decline in public investment, there has been a substantial increase in private investment in groundwater irrigation, particularly in the dry regions of Sri Lanka (Kikuchi *et al.* 2003). Private irrigation investment, which was insignificant in the early 1970s, accounts now for about 20% of the total irrigation investment.

Who benefits from irrigation investment and who should pay for the cost? What is the optimum policy for irrigation financing? Using panel data across 14 states of India during 1970-94, Bhattarai *et al.* (2003) have addressed the questions by separating the direct benefits of irrigation (farmers' share) from the total benefits (economy-wide benefits). The authors estimated the marginal or incremental benefits, both *direct* (farm level) and *total* (rural economy-wide), resulting from irrigation development. Dividing the *total* by the *direct* benefits of irrigation, they computed irrigation *multipliers* for India, which ranged from 3 to 4.5. Using these multipliers, they reached an interesting conclusion that two-thirds or more of the benefits from irrigation development in India are captured by the economy outside of the farm sector. If this is the case, then, there is a need for reconsidering the current policy of trying to recover the costs only from farmers. Clearly, irrigation financing and cost recovery policies are to be designed and assessed on an economy-wide basis. This is an important result from the perspective of both the literature and policy on irrigation financing and cost recovery.

IWMI research has continued to argue for an increased investment in irrigation expansion, especially in small-scale irrigation options and in under-invested areas such as the Sub-Saharan Africa. In regions, such as the Central and South Asia, there is also a need to expand irrigation, not so much on hard options such as the construction of new dams but on soft options such as the development and promotion of small-scale irrigation options that will enhance irrigation access to the resource-poor farmers (see Rijsberman 2003). On the issue of promoting irrigation investment in Sub-Saharan Africa, IWMI has undertaken a major study (IWMI 2005). This study has aimed to address the question of whether the costs of irrigation projects in this region are really high as they are often projected, determine the factors, which influence these costs, and identify cost-reducing options that can make irrigation investments more attractive. For dealing with these issues, this study has analyzed 314 projects implemented during 1967-03 in 50 countries in Africa, Asia and Latin America supported by the World Bank, African Development Bank and the International Fund for Agriculture Development.[7] Some key results of this study can be summarized as follows (Inocencio *et al.* 2005).

(a) If simple regional averages are examined, the unit costs of irrigation projects in Sub-Saharan Africa appear higher than in other regions. But, when the projects are disaggregated to take the average of 'success' projects (those with an internal rate of return of 10% or more), the unit costs of these projects are comparable with those in South Asia, which have the least-cost projects. It is only the very costly and 'failure' projects that drive up the average unit costs of projects in Sub-Saharan Africa.

(b) The profile of key project parameters that characterize a 'success' project suggests that the successful new-construction projects in Sub-Saharan Africa are larger in size (in terms of total irrigated area). These projects were undertaken in more recent years with more farmers contributing to investment costs. They are located in regions with higher rainfall and in countries with a lower per capita of real gross domestic product. The rehabilitation projects characterized as 'success' were also completed in more recent years, but have lesser discrepancy between the actual and planned irrigated areas.

(c) A comparison of the profile between 'success' and 'failure' projects yields two key points. First, the data do not support the argument that project success comes from spending more, but that it depends rather

[7] The sample includes projects with differing purposes (ranging from pure irrigation to multipurpose projects), focus (ranging from new construction to rehabilitation works), types (ranging from river diversion and reservoir-based systems to drainage- and flood-control works), management modes (ranging from state-managed to farmer-managed) and major crops irrigated.

on the simplification and cutting of unit costs. Second, addressing the issue of a relatively high unit cost of irrigation development in Sub-Saharan Africa requires a careful inquiry into the factors leading to a higher probability of failures in the region.

(d) The results of regression analyses suggest that when the factors affecting unit costs are accounted for, there is no statistically significant difference in the average unit costs between the projects in Sub-Saharan Africa and those in South Asia. There is nothing to indicate that the projects in Sub-Saharan Africa are *not inherently more costly* than those in other regions. The message is that when projects are designed to reflect well the characteristics consistent with the unit cost-reducing factors, it is clearly possible to develop and implement projects with lower unit costs.

6.7 MAINSTREAMING GENDER ISSUES

IWMI research has focused on the role of women in irrigation and water management both from performance and empowerment perspectives. Gender studies at IWMI began in a modest way in 1993 with literature reviews and specific case studies in Bangladesh, Nepal, Sri Lanka and Burkino Faso to assess the role of women in irrigation management.[8] These studies generated very important insights into women's contribution to irrigated agriculture. Two early studies, one in Nepal and the other in Burkina Faso, have demonstrated the potential significance of gender analysis in irrigation and water management. Zwarteveen and Neupane (1996) report a case study of the Chhattis Mauja Irrigation System, which is a large farmer-managed system in the terai region of Nepal. This study focuses on the intra-household organization of production to identify who the water users are, how men and women participate in water user organizations and what the implications for system performance are. Despite a high degree of women's involvement in managing irrigated farms, they do not participate in management organizations. Interestingly, such a lack of participation of female water users, rather than being a disadvantage to women, actually enables them to become free riders.[9]

The other study in the Dakiri Irrigation System in Burkina Faso (Zwarteveen 1997) has analyzed the impact of women receiving an irrigated plot on the

---

[8] Results of the earlier research on the gender issues in IMT are summarized in Merrey 1997.

[9] Thus, female-headed households pay less irrigation fees but they are the first to receive water and are not penalized for water theft. This is perhaps an exception very specific to the study site in Nepal (see Zwarteveen and Neupane 1996).

agricultural productivity, labor contribution patterns and intra-household distribution of agricultural incomes. The results show that the labor productivity in women-managed plots is more than twice that of the same plots managed by men. Further, in households with female plot holders, women are able to contribute more to the household's welfare. Having an individual plot sharply increases the efficiency of labor use and clearly enhances women's economic well-being, contribution to their household and their bargaining position. In later years, IWMI research on gender issues has broadened the geographic coverage to other areas and has also considered the role of women in the nonirrigation sectors as well. Buechler and Zapata (2000) have brought together a set of interesting papers on the role of women in the irrigation sector of Mexico. These papers cover the experience of women's involvement in irrigation management in the Bajío and the Lagunera regions and bring the key roles of women in an agricultural setting dominated by a traditionally strong patriarchal structure.

Recent studies on the subject by IWMI researchers highlight the empowerment and poverty alleviation roles of water. Access to irrigation and other production resources is considered an effective way for poverty alleviation among, and empowerment of, poor women. Existing customary practices and social norms, as for example, in most parts of South Asia, do not allow women an equal and direct access to irrigation water. Likewise, some irrigation agencies tend to exclude women categorically from access to irrigation water, though recently some agencies have developed approaches that are based on a sound understanding of the prevailing irrigation-gender relations. There is a need to include both poor men and women stakeholders from an early stage in the planning process for infrastructural development, defining access rules and establishment of water user associations (van Koppen 2000).

In a similar vein, Upadhyay (2003) argues that water polices and programs have proven detrimental to women's water rights. Irrigation interventions consistently fail to recognize the differential property rights, the division of labor and incomes between men and women farmers. IMT programs in many countries require landownership as criteria for membership of water user associations. In most Asian and African societies, the ownership of land is vested with males and this effectively disqualifies women from gaining membership in user associations. There is little evidence for the adoption of any effective approach to redress such problems (Upadhyay 2003).

Although there is increasing awareness of gender issues, there is still a considerable gap between positive intentions of resource managers and concrete action at the field level. An important, but hitherto ignored, reason for this problem is the lack of adequate generic concepts and tools that are policy-relevant and can accommodate the vast variation in irrigation contexts worldwide. The Gender Performance Indicator for Irrigation developed by

IWMI aims to fill this gap (van Koppen 2002). This sociological tool diagnoses the 'gendered' organization to see whether there is any gender-based exclusion in irrigation institutions. The Indicator was applied and tested in nine case studies in Africa and Asia (van Koppen 2002). With these and other studies, IWMI has not only contributed to raising awareness on gender issues in water management but also to developing methodological tools and empirical insights on the subject. Exploring ways to ensure gender equity in water management will continue to be a key feature in IWMI's research agenda in the foreseeable future.

6.8 CONCLUSIONS AND IMPLICATIONS

Water Resources Institutions and Policies remain a major research theme at IWMI. Over the years, the theme has evolved from an almost exclusive focus on institutions for irrigation management to cover a wide range of policy and institutional issues relating to the development and management of water resources at the national and basin levels. Economic and social issues including poverty, gender and equity questions in water resources use and management have been also at the forefront of research in recent years. IWMI research on irrigation management transfer, institutional reforms and poverty analysis continues to remain very influential in guiding national and global water policies.

The analytical and methodological works on basin institutions and water institutional reforms are very useful as a foundation for future research on water institutions. Despite the handful of IWMI studies on water pricing, their contributions to the pricing debate are substantial, in terms of both empirical insights as well as theoretical hypotheses. The research program on gender issues in irrigation development has been one of the pioneering efforts to set the global research agenda on the subject.

In recent years, the comparative analyses of water-poverty linkages, particularly the impacts of irrigation development on poverty and rural livelihoods, have been much appreciated by researchers, policymakers and donor agencies. It is now clearly established that the mere provision of irrigation facilities may have a larger impact on poverty in the initial stages, but to sustain a broader flow of benefits over time complementary policies and institutional reforms are needed. The recent research on irrigation investments in Africa has been instrumental in dispelling the popularly held wrong perceptions about the high costs of irrigation development in Africa. It argues for a more disaggregated analysis of investment possibilities as well as for the importance of project design and its consistency with the factors that can reduce costs such as the size and purpose of projects. On the cost recovery debate, IWMI research has also added a new dimension with its study on irrigation multipliers. Given the fact that the major share of irrigation benefits occurs outside the farm sector, cost recovery policies should not focus exclusively on irrigation water pricing, but need to consider options for capturing revenues from the economy outside the farm sector.

While there is a clear need for irrigation investment in regions such as Africa, there is somewhat a disproportionate attention only on the hard options such as the construction of new dams or rehabilitation of major systems. Instead, IWMI calls for a major expansion of investments in soft options-not just in Asia but also in Africa-such as the development and promotion of water-saving technologies and small-scale irrigation options that will enhance water productivity and access to irrigation, especially among the rural groups that have been bypassed so far by major water-development schemes.

Considering the strategic importance of the research on water institutions and policies, IWMI has elevated this theme as a cross-cutting one anchored in all the four new themes under its new strategic plan (see chapter 2). Now, the research on institutional and policy issues is dispersed across the themes but with a major focus on the institutional and policy assessment of development interventions that are needed to improve water productivity at the basin and subbasin levels. At the same time, works on water-poverty mapping, gender analysis, and policy options for up-scaling innovative practices for basin management, water productivity, and soil and water conservation are also getting an additional thrust and direction under the new thematic structure of IWMI's research.

7

WATER, HEALTH AND ENVIRONMENT

Felix P. Amerasinghe

7.1 INTRODUCTION

Water is the source of life. It is also a source of disease and death. The agents are many and varied, ranging from toxic chemicals to viruses, bacteria, protozoans and helminth parasites, and the human cost is indeed high. Overall, water-related diseases affect some 2.3 billion people and result in 4-6 million deaths each year (Hinrichsen *et al.* 1997). According to World Health Organization (http://www.who.int), diarrhea and malaria cause 22-25% of mortality in children under 5 years of age. An efficient irrigation system should provide greater water security over a longer period resulting in health benefits from improved incomes that translate into more nutritious diets, better housing and clothing, and greater access to and capacity for health care (PEEM 1991). Unfortunately, agricultural systems, especially irrigated ones, have long been associated with manifestations of extreme human ill-health arising from water-related diseases (Hunter *et al.* 1993; Jobin 1999). The major reason is that public health and disease control programs have not been concerns of the water

 More Crop per Drop. Edited by M.A. Giordano, F.R. Rijsberman, R. Maria Saleth. ISBN: 9781843391128. Published by IWA Publishing, London, UK.

resources sector, which typically has focused on potential economic benefits in terms of food production and power generation.

Another concern that has acquired increasing visibility over the past few decades is the extensive exploitation of natural resources to house and feed rapidly expanding human populations during the past 100 years, i.e., the health of the environment that sustains human populations. The destruction of natural habitats, loss of biodiversity, overexploitation of land and water resources, and pollution from anthropogenic sources - all these have contributed to the degradation of the environment in and around human habitations. Agriculture and irrigation, in particular, have been singled out as significant contributors in all of these aspects. Inevitably, such large-scale degradation would threaten the very sustainability of agricultural systems themselves. As McNeely and Sherr (2001) aptly note, environmentalists and agriculturists "will have to recognize that endangered species, essential farmlands and desperately poor humans often occupy the same ground". As in the case of human health, environmental concerns have been neglected in the rush towards rapid development. Yet, unless the key issues such as combating disease, reducing childhood mortality, promoting gender equality and ensuring environmental sustainability that are identified in the Millennium Development Goals (MDGs) are addressed, the primary task of eradicating extreme poverty and hunger in the world (also the primary goal of the CGIAR) will not be achieved. It is this health and environmental dimension of the development challenge that provides the policy context for IWMI research under the theme of Water, Health and Environment (WHE). The general proposition that governs the research activities under this theme is based on the logical link between human and environmental health and its critical implications for the sustainability of food, income and gender benefits.

7.2 RESEARCH EVOLUTION AND APPROACH

Although the WHE theme is relatively new in terms of IIMI/IWMI's history, the attention to health issues has been a part of the Institute's research agenda since its inception. In fact, one of the first international meetings hosted by the International Irrigation Management Institute (IIMI, the forerunner of IWMI) at its original headquarters at Digana, Sri Lanka, was a Workshop on Irrigation and Vector Borne Disease Transmission organized in 1985 by the Joint WHO/FAO/UNEP Panel of Experts on Environmental Management (PEEM), the South Asia Cooperative Environment Program (SACEP) and IIMI. This workshop focused on the links between irrigation and vector-borne diseases such as malaria and Japanese encephalitis, with special emphasis on the 125,000 ha extent Accelerated Mahaweli Development Project of Sri Lanka. It was suggested at the workshop that the links between irrigation and human health were legitimate subjects for IIMI research. However, while recognizing the relevance of the issue, the Institute's position at the time was that it did not have the capacity to actively pursue such a course of action.

However, continuing interactions with PEEM on irrigation and human health issues culminated in the designation of IIMI as an official collaborating center of PEEM in 1987. Subsequently, in August 1991, a PEEM-sponsored consultancy team to IIMI-HQ in Sri Lanka and its country offices in Pakistan and Nepal, as well as to India, recommended health-related research as a potentially important activity for IIMI, highlighting opportunities for incorporating health components into some of IIMI's then ongoing projects. Subsequently, a DANIDA-sponsored Associate Professional Officer (APO) initiated health-related research relevant to irrigation in January 1994. A DANIDA-sponsored workshop in 1997 in association with the Danish Bilharziasis Laboratory (DBL) further explored the irrigation-health research agenda of relevance to IWMI (Konradsen and van der Hoek 1998).

The success of this initiative resulted in human health being formally incorporated into the mainstream of IIMI research through the creation of a separate research program labeled the Health and Environment (H&E) Program. The 'Environment' dimension was included in recognition of the increasing importance of environmental issues in water management. At the time, the Institute did not have significant staff capacity to launch serious research initiatives in this field (the focus was on human health). However, IIMI did have previous research to its credit, focusing on watershed management (such as the Shared Control of Natural Resources project in Sri Lanka) and the environmental aspects of salinity, alkalinity and waterlogging (e.g., Kijne and Kuper 1995) - these were still seen as prime environmental issues linked to irrigation development. The H&E Program purposely left these issues to other IIMI research programs investigating watersheds, irrigation surface water management, groundwater management and systems operations at the time. The H&E program appears to have been originally conceived from separate irrigation-human health and irrigation-ecosystems perspectives, and not from a more holistic 'environmental health' viewpoint that encompassed both humans and natural ecosystems. Thus, these two components have functioned more or less as separate subthematic areas within the program, with only occasional attempts to bring the two strands together (e.g., Steele *et al.* 1997; Harmancioglu *et al.* 2001).

7.2.1 Defining the research context and agenda

When the research on health and environment formally commenced at IWMI, there was only limited recognition of the importance of human health and sustainable ecosystems within the irrigation and agriculture sectors, and no cross-sectoral institutions to actively promote a research agenda investigating such issues. Realizing the direct implications of this research vacuum to its broader water and land mandate, IWMI felt that it could contribute to research and policy in this critical area by:

(a) Putting human health and the environment on the international agricultural development agenda as an important component of rural poverty alleviation and sustainable livelihoods.
(b) Investigating the links between irrigation and health, and developing management strategies to ameliorate some of the biggest rural health problems in developing countries: vector-borne infections (e.g., malaria, Japanese encephalitis, and schistosomiasis), fecal-orally transmitted diseases and illness due to toxic chemicals in water. The hypothesis was that the positive livelihood benefits of agricultural development could be further capitalized by managing irrigation water in a manner that does not contribute to an increased disease burden.
(c) Developing strategies that contribute to the conservation and sustainable use of freshwater-dependent ecosystems within and around agricultural production systems, the hypothesis being that water and land can be managed in a manner that sustains natural ecosystems in river basins, so that the integrity of the environment can be maintained and the flow of natural goods and services ensured.

The Institute's comparative advantages in this endeavor were its research strengths in the agricultural engineering sciences; in policy/institutional aspects relating to agriculture and irrigation management; its links to both national and international partners in the agriculture and water sectors globally (especially in developing countries); a core group of health/environment specialists; and its culture of encouraging cross-sectoral research approaches, bringing to bear expertise in health, engineering, hydrology, agriculture, policy and information-technology resources within a single institution and the access to operating systems, field sites in different regions and countries, and the possibilities for making comparative studies.

From the outset of the research, a multidisciplinary staff has supported the WHE Theme, cutting across the agricultural engineering, health and environmental divide. Interestingly, however, active links with other research themes within the Institute were slow to develop, one possible reason being the time needed for the importance of health and environmental issues to be recognized even internally, and such concerns incorporated into ongoing and new projects whose main focus rightly lay elsewhere. Another reason was the heavy epidemiological emphasis of early WHE health projects-a necessary phase of development in order to establish the credentials of the group and Institute within the biomedical sector. On the environmental side, however, there were links with researchers in other themes, but this aspect of theme work was less pronounced during the early years (a reflection of the in-house expertise available at the time), and it is the health research that took precedence. At present, the WHE theme has shared research projects with the CA in the areas of ecosystems and wastewater research, and with the Water Resources, Institutions and Policies Theme through wastewater and ecosystems research. There is scope for joint research with the Integrated Water Management in Agriculture, Groundwater, and Smallholders Themes in the areas of multiple uses of water, wastewater agriculture, and the human health and environmental impacts of groundwater use, but this potential is yet to be realized.

Specific focus areas for WHE research that evolved during the period 1994-04 were malaria, wastewater-related urban and peri-urban agriculture, multiple uses of irrigation water, pesticides and ecosystems. A synthesis of the research done in these areas is presented below.

7.2.2 Approach and methodology

From its inception, research within the WHE Theme focused on the collection and analysis of empirical, field-based information aimed at establishing agriculture-water-disease and agriculture-water-ecosystem linkages. This has resulted in the development and analysis of large data sets, and qualitative and quantitative outputs based on them. In addition, GIS/RS-based analytical tools and computer modeling were used to further capitalize on the data sets collected. Secondary information-based research activities have supplemented field research.

WHE research has usually involved multidisciplinary teams operating at stakeholder level in farming systems, normally in partnership (be it active or passive) with relevant government agencies and others. More recently, information-technology-based computer modeling and scenario development research have gained more prominence. Overall, the model followed has been to commence with small-scale field/system-based projects and use the results to move towards more generic and policy-oriented considerations. As could be expected, often one field project has resulted in observations that have fostered new hypotheses in a broader multidisciplinary context, and have led to new pilot activities. The key findings and concepts developed within the subthematic areas have come about primarily through empirical field-based interdisciplinary observations. In this regard, proximity to the field has been a key to WHE's success.

7.3 MALARIA AND IRRIGATED AGRICULTURE

Arthropod vector-borne infections have been among the most devastating diseases to afflict humankind throughout recorded history. Many, especially the mosquito-borne diseases, have been associated with water. Malaria takes pride of place among them. About 40% of the world's population (i.e., 2,400 million) is at risk. A minimum of between 700,000 and 2.7 million persons are estimated to die yearly from malaria. Over 75% of these deaths are reported among African children, where between 400 and 900 million acute febrile episodes occur yearly in African children under 5 years of age living in endemic areas (Breman 2001). Overall, 90% of the Global Burden is in Sub-Saharan Africa where the disease is associated with economic losses estimated at up to $12 billion annually and slows economic development by 1.3% per capita per year (WHO 2000; Gallup and Sachs 2001). Malaria has been linked with agricultural development since historical times, and these associations are extensively documented and reviewed (Bradley 1977; Mather and That 1984; Service 1984, 1989; Lacey and Lacey 1990). Recent studies in Africa and Asia have shown that the malaria-agriculture linkages are complex and situation-specific, with greater or lesser malaria prevalence depending upon local conditions and vectors (reviews by Ijumba and Lindsay 2001; Amerasinghe 2004). Research on malaria is global in scale, with multimillion-dollar annual investments, thousands of researchers in diverse fields, hundreds of projects and published papers each year.

Within this massive constellation, however, arguably IWMI has been the leading international agricultural research center involved in malaria research over the past decade, investigating specifically the water-agriculture-livelihoods dimensions of the disease. An Internet search revealed that at least $300 million worth of funding has been made available for malaria research in the past 4 years, primarily in the areas of drugs, vaccines, and genetically modified anopheles mosquitoes that are incapable of harboring the malaria parasite. Unfortunately, there seems to be little recognition of the need for research funding to address issues relating to the agro-ecological, economic and social dimensions of the disease that are on-the-ground realities and continue to bedevil the long-term sustainability of malaria control.

IWMI research on malaria has gradually evolved in approach from 1994 to 2004 from a purely irrigation-malaria nexus relating to Asia to a more integrated water-land-people-based agricultural system management perspective, with emphasis on the African situation. The overarching hypothesis is that malaria-agriculture linkages lend themselves to agroecosystem management interventions that can contribute to a substantial reduction of disease incidence, especially in areas of unstable transmission. A multitude of issues surrounding this hypothesis can be aggregated into three key research questions: How is malaria linked to irrigated agriculture? What are the direct and indirect costs of the disease? What solutions can the agriculture sector offer? These issues have been addressed to greater or lesser degrees over the past 8 years of program evolution. A brief synopsis of the IWMI research done to address them is presented below.

7.3.1 How is malaria linked to irrigated agriculture?

IWMI studies, focused primarily on three Asian countries with the same major vector species (*Anopheles culicifacies*), illustrated the complexity and situation-specific nature of the irrigation-malaria links. The studies portrayed different types of irrigation systems: a traditional small cascade-tank system that is characteristic of rural Sri Lanka (Upper Yan Oya watershed, North Central Province), a medium-scale canal system in India (Kheda district, Gujarat), and a large-scale canal system in Pakistan (Indus Basin Irrigation Scheme [IBIS], South Punjab). In keeping with other Indian studies (e.g., Yadav *et al.* 1989; Sharma *et al.* 1991; Tyagi 2002), rainfall and irrigation water releases were related to population peaks of the malaria vector in the Gujarat case (Konradsen *et al.* 1998a). In the Sri Lankan case study, by contrast, dry climatic conditions, the scheduling of irrigation water releases between the two large irrigation tanks, and water leakage through broken tank sluices contributed to the maintenance of vector-breeding pools within a natural stream that also doubled as an irrigation conveyance channel (Amerasinghe *et al.* 1997, 1999, 2001; Konradsen *et al.* 2000a). The proximity of houses and villages to this primary source of vector production, and the type of house construction were other major risk factors for the disease (Konradsen *et al.* 2003a; van der Hoek *et al.* 1998a, 2003a).

In the southern Pakistani Punjab, where waterlogging is an important issue, an IWMI study did not find a conclusive relationship between depth to groundwater and malaria prevalence, mainly because of the poor reliability of health statistics resulting from low attendance at state clinics (Donnelly *et al.* 1997a,b). Later entomological studies showed that mosquito vector generation in both domestic and agricultural locations in this arid area was primarily dependent on irrigation water, viz. irrigated and waterlogged fields, and communal drinking water tanks (Herrel *et al.* 2001, 2004). From a long-term perspective, however, the irrigation development of the Punjab has been associated with a decline in malaria. The area was highly malarious and epidemic-prone historically, but prevalence is now at a low ebb and the last great epidemic dates back to 1972 (de Zulueta *et al.* 1980). A recent retrospective analysis of the period 1970-99 suggests that the increased abundance of rural *An. stephensi* (a poor vector) relative to *An. culicifacies* (the main vector) may be the cause of the low malaria levels at present. The shift may have been due to waterlogging with related salinization that has created an environment favorable for the more salt-tolerant *An. stephensi* (Klinkenberg *et al.* 2004).

The use of GIS and remote sensing tools has been used to generate malaria risk models in Africa (MARA/ARMA 1998; Kleinschmidt *et al.* 2001). IWMI's foray into the use of such tools in malaria was inspired by the possibilities of manipulating malaria incidence and other secondary data to analyze risk factors at irrigation-system or river-basin scales. Two GIS-based Asian studies within well-established irrigation systems have indicated that in the long term, irrigated areas may not be significantly more malarious than adjacent nonirrigated areas. The first, in the Mahi-Kadana Irrigation Scheme in Gujarat, India analyzed secondary data relating to malaria incidence in relation to water-related factors such as rainfall, rice cultivation intensity, depth to groundwater and irrigation density (Mutuwatte *et al.* 1997). The overall outcome was that irrigated areas had only marginally higher malaria incidence than nonirrigated areas. Depth to groundwater and irrigation density did not explain the pattern of malaria transmission, while the impact of rainfall and rice intensity was inconsistent, being significant in some years and not in others. One rather obvious limitation of this study was its focus only on water-agriculture-climate-related factors, to the exclusion of socioeconomic factors.

The second GIS-based study, in the Uda Walawe Irrigation Scheme in Sri Lanka, included socioeconomic parameters in addition to meteorology, land use, irrigation and malaria–control-related factors. The results, again, were counterintuitive: over a 10-year period (1991-00), the irrigated areas had less malaria than adjacent nonirrigated slash-and-burn agricultural areas in this dry-semiarid environment (Klinkenberg *et al.* 2003a). Rainfall (which generates vector-breeding habitats) was an important independent risk factor at all levels of risk, but factors such as food stamps (a proxy socioeconomic indicator), the extent of forest, and the occurrence of abandoned irrigation tanks were associated with high malaria risk in a multivariate analysis. Interestingly, the extent of paddy cultivation and livestock (cattle, buffalo) husbandry were not significantly associated with malaria risk. The overall message was that in the long term, the functional rice irrigation scheme generated less malaria than the adjacent forested slash-and-burn areas of abandoned ancient irrigation land with semi-functional irrigation structures. Such a conclusion cannot automatically be applied to all irrigation systems: both in Asia and Africa some systems consistently generate more (or less) malaria than others (Amerasinghe 2004), and one of the current research challenges in malaria epidemiology is to precisely identify the determinants.

In addition to rural malaria, Asian countries, such as India, are also familiar with urban malaria transmitted by a vector that breeds successfully in overhead water tanks and other clean water collections in cities (Kumar 1997). It is less well documented in Africa, where the disease is associated with rural communities. For example, malaria parasite prevalence has been estimated to range from 2 to 45 times higher in rural areas than in urban or peri-urban areas in countries such as Ghana, Gambia and Zambia (Gardiner *et al.* 1984; Lindsay *et al.* 1990; Watts *et al.* 1990). None of these cited studies, however, have taken cognizance of the possible impacts of urban agriculture on malaria transmission within the cities themselves. Recent IWMI research has focused on two cities (Accra and Kumasi) in Ghana where an 'informal' irrigation sector cultivates mainly vegetable crops using water from streams and city drains. These studies have provided preliminary evidence of significantly higher vector densities, infective bites, and juvenile parasitaemia rates in communities close to urban agriculture sites than those without such sites (Afrane *et al.* 2003; Klinkenberg *et al.* 2003b). Further studies are in progress to determine whether these higher vector densities and transmission indices are specifically related to habitats created by urban agriculture or to the general terrain in which these urban crops are grown.

7.3.2 What are the direct and indirect costs of the disease?

IWMI evaluations of these socioeconomic aspects of malaria in terms of costs to the affected people, and the costs of national malaria control efforts to the government were limited to the Upper Yan Oya watershed in Sri Lanka. Direct costs averaged 1% of net family income per episode of malaria, while families with multiple episodes spent up to 10% of income per annum. The indirect costs (loss of agricultural work days, labor substitution costs, etc.) averaged 6% of family income per year (Konradsen *et al.* 1997a, b). From the standpoint of national costs for malaria control, a comparative analysis indicated that where feasible, vector control through water management would be the cheapest option, with larviciding of breeding habitats the next cheapest. Other methods such as indoor residual spraying, treatment at hospital clinics, local-level treatment centers, and mobile clinics were more expensive (Konradsen *et al.* 1999, 2000a, b).

Estimates of household costs for Africa are higher. According to the Center for Study of Responsive Law (www.csrl.org), a very low-income African family spends, on average, 28% of its annual household income of $68 to treat malaria alone. Workers suffering a malaria bout can be incapacitated for 5 to 20 days. As a result of this, malaria-afflicted families can harvest, on average, only 40% of the crops harvested by healthy families (www.netmarkafrica.org). For many subsistence families, such a burden could result in severe consequences in terms of loss of food production, farm income and indebtedness. Conversely, reductions in malaria incidence can lead to tangible economic benefits both at the household and national levels. For instance, an interesting study from Vietnam (Laxminarayan 2004) has estimated that the 60% reduction in malaria incidence actually achieved during 1993-98 would have translated to an annual economic benefit of about $10 million and a reduction in the out-of-pocket health care expenses of the households to the tune of about $14 million. Notably, these benefits and cost savings far exceed the total cost of malaria-control programs in the country. An extrapolation of these results implies a 1.8% increase in annual household consumption resulting in an average direct economic benefit of $12.60 for each household in Vietnam.

7.3.3 What solutions can the agriculture sector offer?

IWMI experiences in this area have been mixed, but nevertheless provided valuable insights into the practicalities of implementing an approach to disease control that is often advocated by the health sector but treated with great circumspection by water managers. The studies at the Upper Yan Oya watershed in Sri Lanka provide a case in point. The clear linkage between malaria, major vector-population dynamics and irrigation-water dynamics provided the possibility that irrigation water releases could be managed so as to minimize vector breeding. This feasibility was confirmed by detailed water balance and modeling studies (Konradsen *et al.* 1998b; Matsuno *et al.* 1999). However, subsequent upstream water diversions resulted in insufficient water for the proposed management regimen to be implemented. Instead, another environmental intervention was implemented: the clearing of fallen trees and overhanging vegetation that obstructed the waterway (without damaging the canopy vegetation), the removal of natural obstructions such as rocks, the flattening of the bed, and the consolidation of the embankments in order to facilitate water flow and reduce pooling. In order to reduce ecosystem damage, the intervention was limited to a key malariogenic 7 km stretch of the 20 km waterway. There has been no significant vector breeding in this channel and almost no malaria cases in the area during a 2-year post-intervention monitoring period (Boelee *et al.* 2003). For the irrigation managers, one of the unintended benefits of the intervention was the increased speed of water conveyance, which provided them with greater flexibility in water scheduling.

Another approach to mosquito vector control involves the management of water at field-crop level, especially in flooded rice fields. Originally tested in Portugal in the 1940s, this technique of intermittently wetting and drying rice fields (known as alternate wet and dry irrigation, or AWDI) purportedly resulted in heavy mortality of mosquito larvae and, at the same time, improved rice yields as well. The technique has been tested in North America, Philippines, Japan, China and India with mixed success in relation to the two key factors of mosquito reduction and rice yield improvement (see reviews by van der Hoek *et al.* 2001a and Keiser *et al.* 2002). Except in Japan and China where large-scale implementation has been possible under well-managed irrigation conditions, other studies have been done in experimental plots. Hardly any attention has been paid to AWDI in Africa. IWMI studies, therefore, focused on two areas: testing out the technique under farmer-managed conditions in South Asia (where mainly vectors of Japanese encephalitis breed in irrigated rice fields), and under experimental conditions in Africa (where major malaria vectors occur in the fields). In both studies AWDI saved water locally whilst maintaining rice yields on par with the other water management regimes tested, but there were no significant overall differences in immature mosquito abundance between AWDI and other fields (Mutero *et al.* 2000; Krishnasamy *et al.* 2003). The Indian study, in particular, highlighted the limitations that could be encountered in large-scale farmer-managed field implementation: the leveling of fields was inadequate for proper drainage, and the interval between wettings and dryings was too short for effective kill of stranded mosquito larvae. Rainfall was an added complication that negated the impact of periodic drying. Clearly, the IWMI experiences in Kenya and India (together with some previous studies by others) suggest that the promotion of AWDI as a panacea for rice field mosquito control is premature; it is certainly inappropriate for wet season rice cultivation, and its practicality under farmer-managed conditions still remains to be proven.

Mosquitoes and malaria thrive in poverty-afflicted communities that also are affected by other environmental and economic problems such as malnutrition, poor living conditions, lack of medical care, lack of access to safe drinking water, inadequate household sanitation and waste disposal, food contamination with pathogens, and occupational injury hazards (Mutero *et al.* 2001). The disease requires a holistic approach to unravel the complex interactions between parasite, vector, host, society and ecosystem, and this is provided by the agroecosystem management concept (Forget and Lebel 2001). This takes account of the role in disease burden of factors such as age, gender, education, occupation, family size, nutritional status, location, water management practices, cropping systems, livestock production systems and the effectiveness of public health institutions, in addition to the traditional biomedical epidemiological approaches. Thus IWMI and partners have launched a multidisciplinary, participatory, community-based research project in the Mwea Irrigation Scheme in Kenya to improve the health and economic well being of irrigated riceland communities through researching agroecosystem management practices with the potential to reduce malaria.

Phase-I of the project demonstrated that the Mwea villages exhibited a classic malaria 'paddies paradox' (Ijumba and Lindsay 2001), with irrigated villages having 30-300-fold greater mosquito vectors but two to sixfold less human malaria cases than nonirrigated villages. The reason appeared to be that the major vector, *An. arabiensis*, fed mainly (85-96% of meals) on cattle and other nonhuman hosts in irrigated villages, in contrast to much heavier human feeding rates (42-45% of meals) in the nonirrigated villages (Mutero *et al.* 2003a). Another finding was that experimental rice field pools seeded with ammonium sulfate, a common broadcast fertilizer in paddy rice in the area, generated significantly higher populations of *An. arabiensis* than untreated pools (Mutero *et al.* 2003b). The results suggested that in addition to strengthening malaria-control interventions such as insecticide-treated bed nets, other measures such as zooprophyllaxis using cattle, and the use of alternative chemical fertilizers that did not promote vector population increases, could be practical options for long-term malaria control in this system.

The overwhelming importance of the malaria mortality and morbidity burden, especially in the African context, and IWMI's leadership in malaria research within the international agricultural research sector, resulted in the CGIAR inviting IWMI to lead an inter-center 'System-Wide Initiative on Malaria and Agriculture' (SIMA). The initiative aims to promote research, capacity building and dissemination of information to increase the understanding on the links between malaria and agriculture, and test innovative interventions to strengthen and/or complement existing malaria control strategies. SIMA was launched in December 2001 after extensive stakeholder review and donor commitments for the start-up phase. SIMA complements recent malaria initiatives focused on drugs, vaccines and health care delivery by creating a community of specialists in crosscutting issues on the interface between malaria and agriculture. Specifically, SIMA seeks to (a) deepen knowledge on the impact on malaria of environmental and livelihood changes imposed by different agricultural systems, so that these systems may be managed in ways that mitigate the impact of the disease; (b) determine the impact of malaria on agricultural production, on the livelihoods of rural farmers that are the backbone of the food production system, and on the economies of poor countries; and (c) test innovative anti-malaria interventions for their effectiveness and feasibility of implementation.

7.4 IRRIGATED AGRICULTURE WITH WASTEWATER

Water used in traditional irrigated agriculture is generally biologically and chemically contaminated to some extent during surface runoff, storage, flow in canals and use in agricultural fields. However, there are special circumstances in which highly polluted urban sewage (often supplemented by industrial effluents) is used in agriculture within and around cities, and commonly referred to as urban and peri-urban agriculture. In other circumstances, agricultural water becomes polluted due to mining wastes released into natural streams or irrigation channels in more rural settings. The conventional water, sanitation and health sector generally views polluted water one-dimensionally, focusing only on human and environmental health implications, and advocating technology-based water treatment solutions as prerequisites to reuse. IWMI accepts that the agricultural use of untreated wastewater is undesirable from a health and environmental viewpoint, but recognizes that it is a livelihood reality in many poor countries that cannot afford the investment and maintenance costs of treatment plants to treat most or all of their urban and industrial effluents. IWMI takes a more balanced view of the agricultural use of polluted water in a developing country context, focusing on both costs and benefits in terms of the health, environmental, food chain and livelihoods implications of the practice.

The hypothesis underlying IWMI's research is that urban wastewater[1] is a resource whose use can be managed so as to maximize economic and livelihood benefits while reducing health and environmental risks. Likewise, in the case of water sources contaminated by heavy metals (especially cadmium), some of which get into the food chain and cause human disease, the premise is that agricultural amelioration methods can be developed to reduce such risks. Some of the key questions that arise in such an exploration are: How widespread is the practice of wastewater use in agriculture? What are the costs and benefits of wastewater use? What is the legal and institutional scenario governing wastewater use? What is the threat posed by heavy metal pollutants of agricultural water?

---

[1] *Urban wastewater* consists of a combination of: (a) domestic effluent consisting of black-water (excreta, urine and associated sludge) and gray-water (kitchen and bathroom wastewater); (b) water from commercial establishments and institutions, including hospitals; (c) industrial effluent; and (d) storm water and other urban runoff.

IWMI field studies on wastewater use have been done in semiarid (Mexico, Pakistan, India) and subtropical (Ghana, Vietnam) locations that represented contrasting scenarios of wastewater use: raw sewage used on agricultural land downstream of Guanajuato City (population 75,000) in Mexico; a similar quality of sewage in peri-urban areas around Haroonabad (population 80,000) and Faisalabad (population 2 million) in Pakistan; a combination of partially treated and raw sewage in an urban to rural swathe of agricultural land downstream of the twin cities of Hyderabad-Secunderabad (population 6 million) along the Musi River in Andhra Pradesh State, India; generally more diluted wastewater from drains and streams in the cities of Kumasi (population 1 million) and Accra (population 3 million) in Ghana; and similarly in Vietnam, where sewage is disposed of in river and stream systems from where it is diverted to agriculture. Additionally, in Vietnam night soil (human feces) is directly used to fertilize crops. Detailed studies on the issue of cadmium contamination of agricultural water have been carried out in Thailand.

7.4.1 How widespread is the practice?

According to UNDP estimates, in 1996 there were already 800 million people involved in urban agriculture worldwide. Much of this urban agriculture involved the use of untreated or partially treated wastewater. While the extent of treated and untreated wastewater use globally has not been accurately determined, data extracted from literature sources show that approximately 350,000 ha in 75 cities rely on direct or undiluted use of wastewater, and another 550,000 ha in 17 cities rely on indirect or diluted use worldwide (van der Hoek 2004). A global estimate of approximately 20 million ha of wastewater-irrigated land is sometimes cited (Abayawardana *et al.* 2001; van der Hoek 2002a), but this is speculative, and not based on detailed country-level estimates. Some information on wastewater agriculture extents has emerged from IWMI studies in different countries. In Ghana, the city of Kumasi alone is surrounded by 11,900 ha of peri-urban agricultural areas using untreated wastewater, compared with less than 9,000 ha of 'formal' irrigation for the entire country (Keraita *et al.* 2002). A preliminary estimate from Vietnam provided a value of approximately 7,000 ha, but this is regarded as an underestimate because it was limited to metropolitan areas only (L. Raschid-Sally, IWMI, personal communication). Along the Musi River in India, an estimated 40,600 ha of agricultural land is irrigated by the sewage-polluted river water within a 50-km radius downstream of the twin cities of Hyderabad-Secunderabad, supporting a complex wastewater agricultural economy that provides livelihoods for some 52,000 people in research sites that include the entire urban stretch along the river, four peri-urban areas and four rural villages

(Buechler and Devi 2003a,b). The most detailed study to date is from Pakistan, where field-verified data from representative, size-stratified city samples provided an estimate of 32,500 ha under untreated wastewater agriculture, involving 19,250 families. Although small when compared with the 16 million ha extent of irrigated land in the IBIS, this limited extent of peri-urban wastewater agriculture was responsible for 26% of the vegetable production in the country (Ensink *et al.* 2003). IWMI is presently carrying out detailed ground-truthed estimates of several countries, as a preliminary to making a global assessment of the extent of wastewater use in agriculture. One of the important conclusions from ongoing research is that the importance of wastewater use needs to be assessed not simply in terms of its physical extent, but in terms of its contribution to national production (especially of vegetables and fodder), and to local (city) and household food security, and the livelihoods that it supports.

7.4.2 What are the costs and benefits?

The most important perceived threat to the health of farmers and consumers from the agricultural use of partially/untreated wastewater is from intestinal infections. In general, IWMI studies have confirmed previous work that bacterial levels in wastewater are many orders of magnitude (fecal coliform counts of 10^5-10^9 per 100 ml) above permissible levels as defined in the WHO Guidelines for irrigation water (10^3 per 100 ml) (Buechler *et al.* 2002; Mensah *et al.* 2001; Scott *et al.* 2000; van der Hoek *et al.* 2002a). However, the actual health impacts of these risks are still poorly defined. In Haroonabad, Pakistan, where wastewater was heavily contaminated with helminth eggs in addition to high fecal coliform counts, exposed farmers had a significantly higher incidence of diarrheal diseases, and prevalence of hookworm and roundworm infections, than non-wastewater farmers (Feenstra *et al.* 2000). However, the worm burdens were low and unlikely to seriously affect the health of the farmers.

Recent work in Faisalabad shows that the prevalence of the main diarrhea-causing agents (*Giardia,* causing giardiasis and *Entamoeba,* causing amoebiasis) is more or less the same in wastewater and non-wastewater farmers (J. Ensink, IWMI, personal communication). In Vietnam, data from secondary sources have indicated that hookworm prevalence was consistently much higher in peri-urban vegetable production areas than in nearby ricelands or other field crop areas. Nonagricultural urban populations were at low risk (van der Hoek *et al.* 2003b). However, in the Guanajuato case study in Mexico, the incidence of intestinal diseases in the wastewater-exposed population was 5 times lower than in the city (Scott *et al.* 2000). This statistic was attributed to underreporting of disease incidence and is probably a correct assessment, but the Mexican and

Pakistan studies drive home the point that while risks are higher, wastewater users and their families do not necessarily carry crisis-level disease burdens.

From an environmental perspective, the results of IWMI studies presented a mixed bag. Parameters such as total nitrogen, phosphorus, potassium, chemical and biological oxygen demand (COD and BOD, respectively) and total dissolved solids (TDS) in surface water and groundwater varied considerably from site to site, exceeding FAO recommended safety levels in some instances (Scott *et al.* 2000; Buechler *et al.* 2002; Feenstra *et al.* 2000a,b; Munir *et al.* 2000; Matsuno *et al.* 2001; van der Hoek *et al.* 2002a; Ensink *et al.* 2002a; Mensah *et al.* 2001). Heavy metals exceeded international standards in peri-urban areas of the Musi River downstream of Hyderabad and were significantly elevated above background levels (although within acceptable limits) in peri-urban areas at Faisalabad (R. Simmons, IWMI, personal communication). Both are industrial cities that release untreated effluents into river and sewerage systems.

The benefits of wastewater agriculture are clearly manifest from all IWMI study localities, and demonstrate the water and land productivity, income and livelihoods-sustaining dimensions of this practice. Scott *et al.* (2000) estimated the gross annual water value of the wastewater used in 140 ha of agricultural land in Mexico to be $252,000, and the estimated gross annual value of the nutrient load to be $18,900 ($135/ha/year). In semiarid Pakistan, cropping intensity, irrigation application, land productivity and gross profit margins were higher for wastewater ($840/ha) than for canal water ($614/ha) farmers (Ensink *et al.* 2002a; van der Hoek *et al.* 2002a). In Hyderabad-Secunderabad, India, net annual income per ha averaged €2,812 ($3,225 at current exchange rates) for fodder grass, €833 ($955) for leafy vegetables, and €470 ($539) for banana, while rental income for para grass was €625/ha/year ($716) (Buechler *et al.* 2002). Overall, 100% of rural and peri-urban wastewater users and 83% of urban users studied had income levels above the state poverty line (Buechler and Devi 2003a,b). Likewise in Ghana, irrigated urban and peri-urban vegetable farming, using polluted water, generated incomes ranging from $500-700 to $2,000-8,000 per family per year (depending on crop type and cropping intensity) that enabled these farmers to leap over the poverty line (at the time of study, approximately $300-380 per year) (Danso *et al.* 2002).

7.4.3 What are the legal and institutional scenarios?

It is estimated that more than 50% of the world's population will live in cities by 2050, hence generating very large volumes of solid and liquid waste. Thus on the one hand, wastewater is a problem of increasing magnitude, while on the other it is potentially a valuable resource of water and fertilizer that has economic implications at the community and national level. Mexico is one of the few countries that allow untreated wastewater irrigation for crops (forage, grains, fruits, vegetables and greens) other than those consumed raw. Wastewater farmers legally cultivate such crops for both market and subsistence purposes, but they sometimes also illegally grow crops that are eaten raw. Legal access to wastewater could be obtained through a Municipal Water and Sanitation Board, or from an irrigation district water user association office, or via a presidential decree (Buechler and Scott 2000).

Other countries in which IWMI research has been conducted show little evidence of existing legal or institutional frameworks that could be used to capitalize on the opportunity presented by wastewater. Studies in Faisalabad, Pakistan, show that there is no legislation to effectively manage environmental resources and control pollution. An act passed in 1993 prohibits the discharge of effluent or waste at levels that exceed the National Environmental Quality Standards, but there are no procedural details or regulatory mechanisms to implement this law (Nazim Ali 2002). There is no legal provision to use semi- or untreated wastewater for agricultural purposes, but the Water and Sanitation Authority (WASA) which does not have the resources to treat all the wastewater, disposes of it on land and in surface water channels. In violation of existing regulations, WASA, in fact, provides wastewater for use in agriculture, recovering some of the costs of wastewater collection and disposal by this means.

The situation in Hyderabad, India, is similar. Laws affecting urban agriculture make no provision for wastewater use. Government institutions deny that there are any benefits associated with wastewater agriculture, and some, like the Department of Urban Agriculture within the Ministry of Agriculture deny its very existence. In contrast, The Urban Farmers Association, a private, informal association of farmers, does acknowledge the practice (Buechler *et al.* 2002), which leads to the question as to who is in touch with reality. In Ghana, the city of Kumasi does not have specific laws to regulate irrigated vegetable farming in the city. The Metropolitan Assembly has bylaws that address environmental sanitation, liquid waste collection and treatment, but the city has no infrastructure for waste treatment and safe disposal. The Accra Metropolitan Assembly has bylaws that prevent crops being watered by effluent from street drainage. Though seldom used, it targets poor farmers who have no alternative but to use polluted water, while the polluters themselves are not penalized-an anomalous situation where the polluters include government institutions such as hospitals, ministries, universities, research institutes and other agencies, together with every city household that is not connected to a sewerage system (Keraita *et al.* 2002).

From the foregoing, it is clear that government agencies in many countries either deny the existence of untreated wastewater use or promulgate legislation which is generally unenforceable and frequently flouted by their own agencies. In a situation where cities have no financial means to invest in the construction and maintenance of treatment and safe disposal infrastructure, waste disposal through agriculture, where health and environmental risks can be contained, is an option that is, in IWMI's view, preferable to direct disposal into water bodies where uncontained risks to downstream water users ensue. There is a clear need for legal and institutional frameworks that are realistic, enforceable and make use of the beneficial opportunities afforded by wastewater use in agriculture. IWMI has developed a framework for analyzing the socioeconomic, public health and environmental impacts of wastewater agriculture in developing countries (Hussain *et al.* 2001). An extension of this approach into the methodological issues surrounding determining of the agricultural, public health, soil, groundwater and property value, and ecological and social impacts of wastewater use is reviewed in Hussain *et al.,* 2002. These two contributions make a sound basis for policymakers to review the issues surrounding wastewater use in their particular city or country domains.

7.4.4 What are the threats of heavy metal pollutants?

An issue of increasing importance in developing countries is the pollution of agricultural water (both urban and rural) by toxic chemicals. The risks of heavy metal pollution in urban wastewater have been briefly dealt with in the preceding section. However, heavy metal pollution of irrigation water does not occur from city sewage and industrial sources alone. For instance, arsenic contamination of groundwater from natural (geological) sources has been reported from countries such as the United States, Argentina, Chile, Mexico and Inner Mongolia, and recently received international notoriety as a result of the massive health and environmental disaster in Bangladesh and West Bengal where arsenic contamination was discovered in more that 50% of the groundwater resources used by 95% of the population (Raschid-Sally 2000). An equally important but less well-known issue is the heavy metal contamination of irrigated agricultural lands, especially rice-based systems, receiving uncontrolled discharges from mining, ore processing and smelting. Research over the past 40 years in Japan and China has identified cadmium (Cd) as being particularly dangerous to human health, the long-term consumption of Cd-contaminated rice resulting in Itai-Itai (a form of osteomalacia) or in proximal tubular renal dysfunction (Hagino 1968; Kido *et al.* 1988, 1990).

Recent IWMI research in an isolated and geophysically localized (2,000 ha) heavy-metal-contaminated area in Thailand has resulted in the development of a generically applicable rapid sampling protocol for accurately identifying and quantifying the spatial Cd distribution risk in rice-based irrigation systems (Simmons and Pongsakul 2002; Pongsakul *et al.* 2002; Chaney *et al.* 2002). In addition, detailed biochemical studies involving Cd, zinc (Zn) and iron (Fe) have shown that rice plants effectively control the uptake of Zn and Fe without a corresponding control of Cd uptake, posing a unique risk to human health in subsistence rice-irrigation areas deficient in Zn and Fe. This has implications beyond the health of exposed farming communities: elevated levels of heavy metals (including Cd) in agro-ecosystems have negative effects on food security and livelihoods by reducing yields (via phytotoxicity) and crop quality (through antagonism between heavy metals and essential macronutrients and micronutrients in both uptake and metabolic pathways). Further, heavy metals significantly reduce soil sustainability by destroying soil biodiversity and biologically driven soil nutrient recycling processes. One of the important aims of IWMI research is to test different amelioration methods, so that a strategy can be developed for managing cadmium-contaminated agricultural soils.

7.5 MULTIPLE USES OF WATER

One of the critical issues facing the twenty-first century is the increasing competition for scarce freshwater resources between agriculture, domestic use, industry and the environment. On a global scale, more that 70% of available freshwater has hitherto been diverted to irrigation for food production, but in the 1990s there has been a shift away from agriculture to other users of water (cities, industries, environmental flows), and even in developing countries, water resources currently allocated to irrigation are under threat of being reallocated elsewhere (van der Hoek *et al.* 1999; Meinzen-Dick and van der Hoek 2001). In response, the efforts of the agriculture sector to improve water productivity in terms of more crop per unit of water can reduce the local availability of this resource. For instance, Konradsen *et al.* (1997c) point out that when the agriculture sector takes measures to diminish water losses, access to water for domestic purposes may be greatly reduced and community health may be adversely affected. The community health perspective was an important driver of IWMI's involvement, but the examination of the broader implications of 'more crop per drop' that multiple-use research entailed was complementary to the agenda of the IWMI research theme on Integrated Water Management for Agriculture (Theme 1).

Apart from water for crops, irrigation systems intentionally or otherwise provide water for domestic consumption, home gardens, livestock and permanent vegetation, and also support other productive uses such as fishing, harvesting of aquatic animals and plants, brick-making, and a host of other enterprises (Meinzen-Dick and van der Hoek 2001). The fact that irrigation water serves many stakeholders is not adequately recognized by water managers. In a situation of increasing water shortages and competition, this lack of understanding results in increasing conflicts between stakeholders, and in human health and environmental problems, all of which contribute to reduce the benefits of irrigated agriculture. The hypothesis, therefore, is that conflicts can be minimized and the benefits of irrigated agriculture increased by taking account of actual usage and users of irrigation water. In the context of irrigation systems and river basins, the key questions are: What are the multiple uses and available water sources? What are the quality and quantity issues involved in domestic water supply, and how do they affect human health? How does irrigation system management affect the availability of domestic water? Detailed IWMI research into the different dimensions surrounding the multiple uses of water has been conducted in the Kirindi Oya (command area of 13,000 ha) and Uda Walawe (command area of 18,000 ha) Irrigation Schemes in southern Sri Lanka and in the Hakra-6R Branch Canal (command area 42,000 ha) of the IBIS in the southern Punjab, Pakistan. A smaller study has been carried out in the Basse Moulouya Irrigation Scheme in northeastern Morocco.

7.5.1 What are the multiple uses and available sources?

Using the Kirindi Oya Irrigation and Settlement Project (KOISP) as a case study, Meizen-Dick and Bakker (1999) and Bakker *et al.* (1999) examined the complex issues surrounding the transfer of water out of agriculture to meet other demands, often without the agreement of, or compensation to, farmers with irrigated land and water rights. They found a range of water uses and users in the system, from field crops to domestic users, home gardens, cattle owners, fishers and wildlife conservation areas, and determined that there were important residential, gender and class differences among the water users. Among important management issues were the allocation of irrigation water during periods of scarcity, the improved management of irrigation tanks in the wet season so as to allow dry-season irrigation, and the importance of water quantity and quality not only for domestic use, but also for livestock, fisheries and wildlife.

Meinzen-Dick and Bakker (2001) established a framework for examining the statutory and customary water rights of multiple users of water in the Kirindi Oya system, demonstrated that stakeholders went far beyond the owners and cultivators of irrigated fields, and established that these diverse groups were claimants on the management of the water resources system and needed to be included in considerations relating to the transfer of water from irrigation to other uses. An important point in this argument is that the other uses are not highly consumptive, and therefore can be accommodated in irrigation design and management. Furthermore, the economic benefits of multiple uses, relating to livestock and fisheries in particular, could be substantial. Renwick (2001), for instance, has estimated that the contribution from fisheries in three reservoirs of the KOISP was about $500,000 per year, which represents an addition of 18% to the value of the annual rice production of KOISP and an important contribution to improved nutrition and reduced poverty.

A review of literature from Asian (e.g., Pakistan, Sri Lanka) and African (e.g., Morocco, Senegal, Cameroon, Egypt, Zimbabwe) countries shows that open canal irrigation may be the only source of water for all uses, especially in arid environments (Boelee *et al.* 2000). In the Pakistani Punjab, the Hakra-6R distributary canal is located on land reclaimed from the Cholistan Desert and has very limited natural water resources. Waterlogging and salinization are serious problems, and groundwater is brackish and unfit for human consumption. All domestic water sources in the area originate from the irrigation system, the most important of which is the *diggi* (village tank), which is filled weekly with water from the irrigation canal. Other sources are seepage water from canals and fields, and direct use from canals. Only 12% of 364 sampled households had access to a public water supply scheme, which was also supplied from the irrigation system.

The main reason for preferring a particular source was the perceived water quality (overwhelmingly, seepage water for drinking and cooking) and distance from the particular use (for other uses such as washing utensils, laundering, bathing and sanitary ablutions) (van der Hoek *et al.* 1999). In the Basse Moulouya Irrigation Scheme, Morocco, with similar high salinity groundwater, irrigation water is stored in covered private or communal tanks. The water is also transported to nearby areas without irrigation at high prices, up to $\$5/m^3$, while farmers buy the same water for $\$0.023/m^3$. The water is used for all kinds of domestic purposes, including drinking, and productive purposes such as watering livestock and tree nurseries (Boelee and Laamrani 2003). The Kirindi Oya Scheme in the southern semiarid zone of Sri Lanka also suffers from brackish groundwater and problematic domestic water supply; so, the irrigation authorities have constructed a piped water supply originating from the main reservoir to the irrigated areas. A sampling of 156 households in this system showed that standpipe and fresh well water were favored for drinking, cooking and washing utensils, and several water sources (standpipes, wells, canals, river and irrigation tank) favored for laundering, bathing and sanitary ablutions (van der Hoek *et al.* 1999). Despite the availability of standpipes and wells, the majority still preferred to use canals, rivers and tanks for bathing and laundering - probably indicative of a cultural preference for flowing water for these activities.

7.5.2 What are the health impacts of water quality/quantity?

Insights into the health implications of domestic water quality and quantity have been gained from a series of IWMI studies in the Hakra 6R system in Pakistan (Ensink *et al.* 2000, 2002b; Feenstra *et al.* 2000a; Jensen *et al.* 2001, 2002, 2003; Nielsen *et al.* 2001, 2003; van der Hoek *et al.* 2001b, 2002b) and the Uda Walawe scheme in Sri Lanka (Meijer 2000; Prado 2002; Rajasooriyar 2003; Shortt 2001; Shortt *et al.* 2003; van der Hoek *et al.* 2003c). In Pakistan, seepage water (from canals, reservoirs, ponds and agricultural fields) was much less bacterially contaminated (on average, 2 *E. coli* colonies/100 ml) than surface water taken directly from canals, reservoirs and, when present, the village water supply scheme (30-830 *E. coli* colonies/100 ml). However, the biological quality of drinking water stored in-house was poor (on average 27 *E. coli* colonies/100 ml), regardless of the external source. In Sri Lanka, the surface water sources were highly contaminated bacteriologically (200-7000 *E. coli* colonies/100 ml) and parasitologically (44 *Giardia* and *Cryptosporidium* oocysts/liter), whereas tube well water was significantly cleaner (17 *E. coli* colonies/100 ml; 07 oocysts/liter).

In Pakistan, simple intervention studies were done to assess whether bacteriological risks could be reduced in-house and at source. One study that compared biological water quality in traditional wide-necked in-house drinking water storage pitchers with experimental narrow-necked pitchers that prevented direct hand contact with the water, showed that the intervention was effective only when the water source was relatively clean (i.e., had <100 *E. coli* colonies per 100 ml), and also that it could not prevent events of extreme pollution at the point of collection (Jensen *et al.* 2002). Another study showed that bacterial counts were 60-fold higher at household taps and fivefold higher in household pitchers in a non-chlorinated village water supply system than in a chlorinated one (Ensink *et al.* 2000). The innovation here was that the chlorination process was paid-for and operated entirely by the village, and not by a water supply agency.

Chemical contamination of irrigation water was also evident from the IWMI studies. In Pakistan, deep wells had high salinity (EC averaging 120 mS/m) whereas water from surface water sources such as seepages, shallow wells and water supply scheme was less saline (EC 20-40 mS/m) (Ensink *et al.* 2002b). In the Uda Walawe irrigation area, too, salinity and fluoride were higher in groundwater (EC 29-69 mS/m, fluoride up to 5.8 mg/l) than surface water (EC 17-25 mS/m, fluoride 0.39 mg/l) (Rajasooriyar 2003; Shortt 2001; Shortt *et al.* 2003). Thus farmer families were at risk from bacteriological or chemical pollutants depending on the domestic water source. There were early indications of fluoride poisoning manifested as dental fluorosis among 43% of sampled school children in the area, and a strong association between water fluoride levels and dental fluorosis (Ekanayake and van der Hoek 2002, 2003). Shallow wells near canals were less contaminated than those further away (van der Hoek *et al.* 2003c), highlighting the importance of seepage water in diluting fluoride levels. This was confirmed by hydrogeology studies and isotope analysis (Prado 2002; Rajasooriyar 2003).

It has previously been accepted that water quality was the prime determinant of water-related diarrheal diseases and that better quality would ensure better health. The general consensus now is that quantity is at least as important, if not more important, than quality for diarrheal disease control (Kolsky 1993; Cairncross 1997). There are no guidelines for domestic water quantities (apart from the minimum value of 25-30 liters/capita/day), and because of this, no standard methodologies for estimating such values. IWMI research in Pakistan developed a methodology for estimating water quantities in the context of the multiple uses of water and the cultural sensibilities of the population, providing two indices for use per capita per day: a) in terms of liters, and b) in terms of number of events of water contact (Ensink *et al.* 2002b). Water use surveys showed large differences in water availability at household level, ranging from highs of 48-113 liters/capita/day for those with pumps and storage tanks to lows of 10-15 liters/capita/day for those without any water connection or storage facility. Observable water contact events (for bathing, and washing clothes and utensils) were greater for those without water pumps or storage (3-4 events/capita/day) than for those with pumps and storage (0 events/capita/day, because such activities occurred within the household and were not observable). Over 70% of households did not have the 50 liters/capita/day considered to be adequate for healthy living conditions (Gleick 1998).

van der Hoek *et al.* (2001b) demonstrated that the association between water quality and diarrhea varied by the level of water availability and the presence or absence of a toilet. Among households with adequate water and a toilet, diarrhea incidence was higher when more *E. coli*-contaminated surface water was used as a drinking water source than when cleaner seepage water was used. For those with less domestic water availability, and using the cleaner but quantitatively less available seepage water diarrhea incidence was higher than for those using the plentiful but contaminated surface water sources. In a multivariate analysis, no direct association was found between water quality and diarrhea incidence, but water quantity, mediated through sanitation and hygiene behavior, had a significant impact in reducing diarrhea.

7.5.3 How does irrigation affect domestic supply?

The operation of irrigation canals had important consequences for the availability of domestic water in the Uda Walawe study. Canal seepage accounted for 74% of groundwater recharge, and canal closure resulted in groundwater levels decreasing by 1-3 meters within a few days, leading to the drying-up of many shallow wells and problems of domestic water access for farmers (Meijer 2000). Concrete lining of canals has further restricted the availability of good-quality water for domestic use, thereby threatening the health of the population (Boelee and van der Hoek 2002). In this instance, the attempt at more efficient irrigation management through canal lining deprives the rural population of their best source of drinking water, and the authors point out the need for inter-sectoral planning and management of water resources to ensure that the needs of at least the most important stakeholder group in the system (farmers) are met.

The IWMI studies in Pakistan and Sri Lanka highlight some of the most serious deficiencies in irrigation management: lack of the recognition and rights of the many uses and users of water in an irrigation system; lack of understanding of the health implications of water availability (in terms of quality and quantity) to meet the basic domestic requirements of the farmers who, after all, are the primary stakeholders in the system and produce the food; and lack of inter-sectoral planning in constructing and operating systems so as to mitigate the conflicts and adverse health and environmental impacts resulting from inadequate water supplies and competition for these supplies. The results from Morocco show that the benefits of irrigation increase substantially if the irrigation managers do take the multiple uses of water into account (Boelee and Laamrani 2003). Bringing these issues to the attention of policymakers and planners is an important responsibility of the Institute in the immediate future.

7.6 HEALTH HAZARDS OF FARM PESTICIDES

Modern intensive irrigated agriculture is dependent on two critical external inputs for its success: chemical fertilizers and pesticides, the latter being toxic chemicals. Farmers are occupationally exposed to these toxic chemicals through handling the pesticides and breathing the spray droplets. Moreover, the ready availability of these chemicals within households and nearby village sales outlets opens the door for accidental or deliberate ingestion of pesticides, which is one of the major causes of acute pesticide poisoning and death in irrigated agricultural systems (van der Hoek *et al.* 1998b). The issues surrounding pesticide use are not directly tied to water, but indirectly, agricultural policies and water management influence crop selection, weed control, etc., which in turn influence the type and intensity of pesticide use.

IWMI investigated direct consumption and occupational exposure to pesticides, the rationale being that pesticides are an important cause of illness and death in farming communities, leading to human suffering, loss of productivity and poverty. The key issues investigated were: (a) What is the extent of the problem of acute pesticide poisoning in irrigated areas? (b) To what extent is lack of knowledge a factor in pesticide poisoning? (c) What are the risks to farmers of occupational exposure to pesticides? (d) How can agricultural management policies help reduce occupational exposure to pesticides? Reported herein are a distillation of results from studies carried out in Pakistan and Sri Lanka.

7.6.1 What is the extent of acute pesticide poisoning?

High rates of accidental and intentional pesticide poisoning occurred in the agricultural communities of the two countries investigated. This was related to the easy availability and regular use of these toxic chemicals. In Pakistan, for instance, pesticide use increased from 665 metric tons in 1980 to nearly 45,000 metric tons in 1997, mainly on cotton, rice and vegetable crops. Interestingly, insecticides were the most heavily used category of pesticides, in contrast to most developed countries where herbicides predominate - the reason very likely being the need for high inputs for insect pest control associated with the cotton crop in Pakistan (Feenstra *et al.* 2000c). This increase in usage was associated with several reported outbreaks of acute pesticide poisoning in 1963, 1972, 1976 and 1984 related to different insecticides. Several hospital studies indicated that 35-77% of cases were due to accidental poisoning and 23-64% to intentional poisoning. In irrigated farms in the Sindh, however, accidental poisoning accounted for 95% of these cases and only 5% were intentional poisoning (Feenstra *et al.* 2000c). Unfortunately, neither this field study, nor the wider literature review, quantified the mortality resulting from pesticide poisoning.

Studies in Sri Lanka showed that 73% of the cases of acute pesticide poisoning in the country were due to intentional poisoning and 27% to accidental or occupational exposure (van der Hoek *et al.* 1998b). IWMI studies showed that within irrigated areas, most cases (83%) were due to intentional poisoning, and 18% of these died. A study in North-Central Sri Lanka, during 1991-94, showed that 59% of 526 pesticide-related hospital admissions were due to organophosphate, 4.5% to carbamate, 8% to organochlorine insecticides, and 28% to herbicides, sulfur and other unspecified chemicals. In southern Sri Lanka an investigation of 242 hospital poisoning cases showed that 20% of cases were due to organophosphate insecticides, 8% to carbamates, 6% to organoclorines, 3% to pyrethroid and pyridine insecticides, and 22% to herbicides and sulphur (van der Hoek and Konradsen 2004). There were important differences in personal and family characteristics between cases and controls: underlying mental disorders such as depression, alcohol dependency, ending an emotional relationship, a history of pesticide poisoning in the family, unemployment, being single, and low educational status came out as the main risk factors for acute pesticide poisoning in this study.

7.6.2 What role does knowledge play in pesticide poisoning?

A 1997 IWMI study of 1,354 farms in an irrigated area of the Sindh, Pakistan, showed that 60% of 1,080 farmers who used pesticides were aware of the health hazards of these chemicals (Feenstra *et al.* 2000c). The Sindh is a region with a poor literacy rate, i.e., 38% for total population and 26% among women in 1998 (www.edumag.com/statsliteracy.html). However, Feenstra *et al.* point out that while it is generally assumed that poor farmers have little knowledge on the proper use and potential hazards of pesticides, studies have provided conflicting evidence, ranging from very poor knowledge to high knowledge among men but poor knowledge among women, which, in fact, reflects the sociocultural and educational situation in Pakistan. Unfortunately, the study by Feenstra *et al.* (2000c) did not assess in detail the farmers' use of protective measures or the reasons for nonuse of these protective measures. They comment, however, that even where there was knowledge, the use of precautionary measures was limited by inadequate spraying equipment and hot climatic conditions that made protective gear uncomfortable to use.

In contrast to Pakistan, Sri Lanka is a country with a high literacy rate (91% total, and 85% among women; www.unesco.org), easy access to print and electronic media, greater rural gender equity and equal educational opportunities for both genders. Yet available figures show equally high rates of pesticide-related problems as in Pakistan: an annual mean of around 13,000 poisoning cases per year (van der Hoek *et al.* 1998b). There was good awareness of the toxic effects of agrochemicals, but little effort to practice safety measures such as wearing gloves, mouthpieces or protective clothing because of the discomfort involved. Farmers were well aware of negative effects of pesticide spraying, but continued to use excessive quantities of agrochemicals because of the perception that frequent application was necessary to protect their crops (van der Hoek *et al.* 1998b). The two studies show clearly that higher awareness of health risks does not automatically translate into behaviors that would reduce exposure. Interestingly, in the more literate Sri Lanka, the proportion of intentional poisonings was far higher than in Pakistan.

7.6.3 What are the risks of pesticide exposure?

The impacts of occupational exposure to pesticides are poorly studied in developing-country irrigation systems. Generally, widely used organophosphate and carbamate pesticides are known to have acute systemic effects, mediated largely through cholinesterase inhibition, that lead first to overstimulation and then to depression of the nervous system. IWMI studies in rural Pakistan have shown pesticide residues in human blood and tissues, as well as evidence of chronic pesticide poisoning due to long-term exposure - the enzyme acetyl cholinesterase (AChE) has been shown to be inhibited to levels <30% of normal in the majority of farming men and women in agricultural areas (Feenstra *et al.* 2000c). In the Sindh, where cotton (69%), rice (18%) and vegetables (10%) are the main cultivated crops, 18.7% of 1,080 farmers using pesticides have reported health problems associated with these chemicals among their family members (Feenstra *et al.* 2000c). In Sri Lanka, an IWMI study of 216 farmers (including a group trained in Integrated Pest Management [IPM] techniques that emphasized safe and minimal use of pesticides) and a control group of 55 fishermen revealed that 24% of farmers had suffered at least once from symptoms of acute occupational pesticide poisoning. Non-IPM farmers were five times more exposed to pesticides than IPM farmers. AChE inhibition during the high exposure season was significantly higher (8-11%) in both farming groups than in the control fisher group (3%), but the level of inhibition was significantly less in the IPM farmers than in the non-IPM farmers. Seven of the 16 acute symptoms used in the study were significantly positively associated with increased AChE inhibition (Smit *et al.* 2003). A clinical investigation of 30 farmers (compared with 30 control fishermen) in the same study area revealed both sensory and motor nervous damage due to the effects of long-term exposure to organophosphate insecticides (Peiris-John *et al.* 2002).

7.6.4 How can agricultural management reduce the risks?

The results on occupational exposure to pesticides in relation to IPM provide a strong impetus to the initiative to widen the implementation of this technique as a sustainable alternative to high agrochemical input agriculture in Sri Lanka. The additional health benefits of lowered pesticide exposure supplement the IPM benefits of increased yields and lowered expenditure on agrichemicals (van den Berg 2002). As a further measure to assist countries regulate against and reduce dependency on toxic agrochemicals, IWMI researchers, in association with other national and international colleagues have called upon relevant bodies such as the WHO and FAO to develop a Minimum Pesticides List that would identify a restricted number of less dangerous pesticides to perform specific tasks within an integrated pest management framework (Eddleston *et al.* 2002) Also, IWMI researchers have provided key inputs to the international network formulating policy recommendations on pesticides with the aim of reducing occupational exposure (Konradsen *et al.* 2003b).

An External Review of the Institute in February 2000 rather controversially recommended that IWMI's health research focus should be confined to topics directly endogenous to water use, and as such, discouraged research on issues such as the health impacts of pesticide use in irrigation systems. This aspect of health research was, therefore, discontinued after the completion of ongoing projects.

7.7 ECOLOGICAL ASPECTS OF AGRICULTURE

The development of IWMI research interests in the ecological aspects of agriculture in river basins arose out of the realization that water and its uses had to be considered in a broader integrated management context (as against a narrow focus on irrigation management alone), and that all aspects of water use (agricultural, domestic, industrial, and environmental) needed to be taken into account in water resources policies. IWMI research during the period under review focused primarily on the impacts of irrigated agricultural development on downstream coastal wetlands, issues surrounding the sustainable use of inland wetlands, the estimation of the environmental water requirements of river basins and issues of biodiversity conservation through the implementation of the principles of eco-agriculture. Most of these are new and ongoing initiatives. The WHE program was initiated in 1998, but in-house capacity in the fields of ecology and environmental science was strengthened only over the past 2-3 years. A recently developed framework for this new research initiative (Smakhtin 2002) is relevant not just for IWMI, but also for other international and national institutes with an interest in, or mandate to implement, the integrated management of water resources.

At its inception, the hypothesis within this subtheme was that the operation of irrigation systems has impacts (be they positive or negative) on downstream wetland systems. This later led to broader considerations of the sustainable use of inland wetland systems. More recently, these and other ideas have been consolidated into a single, more general, argument that tools and methodologies could be developed and successfully applied in developing countries to manage water and land resources in a manner that optimizes agricultural production while conserving freshwater-dependent ecosystems and their biodiversity. The key issues addressed in this subtheme during the period under review were: 1) What are the impacts of irrigated agriculture on natural ecosystems and their biodiversity? 2) How can inland wetlands be sustainably used so that food production and livelihoods are enhanced while preserving wetland integrity? 3) How can environmental water requirements be determined and allocations operationalized within river basins? The results represent the outcome of work done in South Asia and more recently in Sub-Saharan Africa.

7.7.1 What are the ecological impacts of irrigated agriculture?

IWMI's first foray into environmental research in 1997 arose out of its work within the Kirindi Oya Irrigation and Settlement Project (KOISP) in southern Sri Lanka. This resulted in the development of a framework for valuing water for its multiple, and often competing, uses, with a special focus on wetland ecological services and the management implications of valuing water for competing uses in terms of evaluating trade-off scenarios (Bakker and Matsuno 2001). Initial studies showed that at a macro-scale, increased output from the irrigation system was likely offset by loss of ecological value of a downstream RAMSAR wetland system and a consequent decrease in tourism revenues; at a stakeholder level, the economic value and employment generation through increased paddy production were likely offset by declining shrimp and finfish productivity leading to poverty and loss of livelihoods. Subsequent studies showed that this was indeed the case: Higher-value brackish water finfish species were replaced by lower-value freshwater species, and the high-value shrimp fishery crashed (Amerasinghe *et al.* 2002; Jayakody 1993; Kularatne 1999).

The impact on water birds, however, was less than anticipated: bird diversity and numbers were, in fact, higher at the irrigation-affected lagoons than at an unaffected lagoon, and only a few specialized brackish water species appeared to be affected (Amerasinghe *et al.* 2002). Upstream irrigation water in the KOISP was within internationally accepted quality standards, but drainage and lagoon water had values of turbidity, phosphorus and ammonia that exceeded the quality standards for aquatic ecosystems or aquaculture (Matsuno 1999; Matsuno and van der Hoek 2000). The quantity of irrigation outflows entering affected lagoons diluted the salinity of the lagoon water, and resulted in continuous outflows into the sea (Matsuno and van der Hoek 2000). Upstream irrigation canals discharged >6,000 kg of nitrogen and >600 kg of phosphorus per month into these lagoons, thereby contributing to eutrophication (Piyankarage 2002; Piyankarage *et al.* 2002).

Smakhtin and Piyankarage (2003) used a parsimonious modeling approach tailored to suit data-deficient situations to simulate the water-level conditions that likely existed in the lagoons affected by the KIOSP prior to irrigation system development. A daily reservoir model was used, that allowed the pattern of hydrological variability in the lagoons to be reproduced without the need for detailed catchment processes simulations. If ecologically important lagoon water levels are specified in the future (for example, levels needed for successful use of the lagoons by winter migrant wading birds), changes in their frequency can be evaluated using the frequency curves generated by the model, which also permits estimations of surface water area and volume, and long-term records of such parameters can be used to calculate salt concentrations and flooding dynamics. This parsimonious modeling approach has been extended to other situations for example, to assess the potential impact of irrigation water releases from a planned 10,000 ha extension of the Uda Walawe Irrigation Scheme in southern Sri Lanka on the brackish water Karagan Lagoon at the foot of the Walawe River Basin (Stanzel *et al.* 2002); and to simulate a continuous time series of estuarine mouth openings and closures for ungauged lagoon/estuary systems, using the Umgababa Estuary in KwaZulu-Natal, South Africa, as an example (Smakhtin 2004). Although presently tested on only a few coastal lagoons, parsimonious modeling approaches (summarized in Smakhtin *et al.* 2003a) have great potential as practical management tools for fragile coastal ecosystems.

Another aspect of agricultural expansion is the threat to natural resource areas and their wild biodiversity resulting from habitat change. A study commissioned by the World Conservation Union (IUCN) has resulted in the formulation of a set of strategies for sustainable agricultural development under the banner of 'eco-agriculture', whereby agricultural systems are designed and operated so as to sustain as much wild biodiversity as possible, while simultaneously improving livelihoods through increased agricultural productivity and income generation. IWMI is presently in the second year of a 4-year project in a new 10,000 ha irrigation scheme in southern Sri Lanka, carrying out a multidisciplinary longitudinal study on biodiversity changes associated with the transformation from a forested to an irrigated rice-banana ecosystem, together with the implementation of eco-agriculture principles in the design of the new system, and evaluation of the changes in the socioeconomic condition of the settlers over time (Bambaradeniya and Amerasinghe 2002; Amerasinghe *et al.* 2003). The predevelopment biodiversity assessment component is now complete, and the study is continuing at present through the infrastructural construction phase of the irrigation project. It is hoped that the successful completion of this project will provide a model for sustainable agricultural expansion in the Asian context.

7.7.2 How can wetlands be used sustainably?

Inland wetlands are increasingly used for agriculture, and also provide domestic water for humans, livestock and aquaculture, and materials for handicrafts, brick making and fuel. In addition, they are believed to have other hydrological functions such as flood attenuation and maintenance of river base flow that are extremely important to humans and ecosystems. Environmentalists concerned about biological conservation and agriculturalists focused on food production have often worked at cross-purposes in wetlands. With increasing population pressure, human use of wetlands is expected to further increase, and regional and national development agencies and NGOs are now exploring options for intensified and expanded wetland use, while conserving this important natural resource base.

One of the objectives of IWMI research is to quantify wetland hydrological functions. Smakhtin and Batchelor (2004) have investigated the role of inland wetlands in water flow regulation (specifically flood attenuation and base flow regulation) based on continuous observed stream flow records and flow duration curves (FDC) using the Rustenberg wetland in South Africa as a case study. They conclude that FDC a) are a simple and convenient means to analyze different aspects of changing flows if either representative observed flow records exist or reliable flow time series can be simulated; and b) indicate how flows of different magnitude are affected by wetlands, but can also be used for more specific analysis of base flow contribution, hydrograph visualization in years of different wetness, and analysis of continuous flow events above or below predefined threshold discharges, which can have ecological or water-resource implications.

Another objective of the IWMI research program is to develop an integrated framework for participatory technical, agronomic, socioeconomic and institutional interventions for sustainable wetland use with implementing agencies such as government extension departments, community action NGOs and farmers. The focus areas for research include hydrology, biodiversity, management of soils, and actual and potential wetland use. IWMI-led research has already investigated the usage patterns of 10 representative wetlands in South Africa, Swaziland, Zimbabwe and Zambia (Masiyandima *et al.* in preparation). A review of existing wetland classification systems and a framework of biophysical attributes that would define the potential for different types of use are also being developed, in order to provide a basis for evaluating and prioritizing development options for southern African wetlands.

7.7.3 How can ecological water needs be determined and met?

A global assessment has shown that while 40% of the world's populations live under conditions of water stress (and projected to increase to 50% by 2025), human activities have degraded freshwater ecosystems, reduced their capacity to support biodiversity, and reduced their capacity to provide goods and services to humans (Revenga *et al.* 2000). However, the water requirements of aquatic ecosystems have not been explicitly addressed in such assessments, nor have they been estimated globally. As part of IWMI ecosystems research, Smakhtin *et al.* (2002, 2003b) have presented a pilot assessment of global environmental water requirements, based on the simulation of global water use and hydrology data. The conceptual framework characterized the water requirement in terms of low flow and high flow components, both of which are important for the maintenance of ecosystem integrity, and explicitly related to flow variability. Not surprisingly, the study showed that reservations for environmental purposes will decrease the water availability for humans, and that even at modest levels of environmental water requirements, parts of the world may already, or in the near future, be environmentally water-scarce. The methodology is still preliminary-for instance, it is primarily hydrology-based and no ecological indicators are incorporated that would provide more realistic ecosystem water requirement estimations. Temporal and spatial scales also need to be accommodated, so as to allow for seasonal changes and for the development of more detailed regional or river-basin-level assessments.

7.8 CONCLUSIONS

Over the past 10 years, IWMI research on water, health and environment has clearly made significant contributions at multiple scales and in multiple arenas. The WHE Theme, which began as the Health and Environment Program in the mid-1990s, has clearly achieved and exceeded the original objectives of:

a) Putting human health and the environment on the international agricultural development agenda as an important component of rural poverty alleviation and sustainable livelihoods.
b) Contributing to a better understanding of the links between irrigation and health and development of appropriate strategies to ameliorate some of the largest rural health problems in developing countries.
c) Developing strategies that contribute to the conservation and sustainable use of freshwater-dependent ecosystems within and around agricultural production systems.

The resultant outcomes and impacts of the past decade of IWMI research on water, health and environment have not only contributed to the Institute's overall mission of improving the management of water resources for food, livelihoods and nature, but are also directly addressing several of the Millennium Development Goals (MDGs) such as combating malaria and other diseases (e.g., through SIMA and wastewater research); ensuring environmental sustainability (e.g., through agricultural ecosystems research); and developing a global partnership for development (e.g., through SIMA and the wastewater initiative). Indirectly, WHE research also supports other MDGs such as promoting gender equality and reducing childhood mortality through sensitivity to these issues in IWMI's activities. All of these endeavors ultimately contribute to the first MDG of eradicating extreme poverty and hunger.

As IWMI now embarks on a new research paradigm, introduced in chapter 1, the Institute has also reformulated its approach to water, health and environment research. As noted in section 7.2 above, IWMI's research on health and environment has largely been addressed as two separate issues rather than a holistic environmental health program. In addition, health issues are not only linked to the environment but are also a critical component of overall poverty reduction and general livelihoods strategies. Thus, in IWMI's new thematic structure, health, like institutions and policies, has been transformed into a community of practice that is integrated within all four of the new IWMI themes. Health issues will thus be incorporated into water productivity and water poverty analyses as well, and in particular, in assessing the impacts of high-potential interventions to address these two issues.

Environmental issues have been incorporated into the new Theme Water Management and Environment, which concentrates on further progressing many of the key research areas initiated under the WHE theme, namely: addressing environmental water requirements in basins, enhancing benefits of agriculture-wetlands interactions, and valuing contributions of ecosystem services to livelihoods. In addition, just as health, environmental issues are now an integral part of the Institute's overarching research framework which aims to identify and assess the impacts of high-potential interventions to improve water and land productivity, access of the poor to productive water and land resources, and the sustainability of the natural resource base. We believe this new formulation not only builds on the significant accomplishments of the WHE theme but also better integrates health and environmental issues into the Institute's overarching research agenda.

8

Water Management for Agriculture

David Molden

8.1 INTRODUCTION

Over the last decade, research and policy issues on water, food and environment have once again moved to the forefront of global, national and local discussions on sustainable development. This renewed interest reflects new concerns as well as a change in the overall focus of the debate. From the 1950s to the 1980s, the discourse on water and agriculture was largely dominated by development pressures to increase investment in the construction and improvement of irrigation facilities. Since the early 1990s, however, there has been growing concern as to the real impacts of irrigation investments, especially within the international donor and development community.

Irrigation certainly played a key role in fueling the Green Revolution, which in turn led to an unprecedented growth in world food production.[1] However, with comfortable food production, declining water availability and water-related

[1] The growth rate of the world's food production, though declined by 0.4% in the nineties, still substantially outpaces the growth in population. This is especially true in the developing countries, where food production increased 3.4% annually, exceeding the annual population growth of 1.5% in the 1990s.

ISBN: 9781843391128. Published by IWA Publishing, London, UK.

environmental degradation, the justification for additional investment in irrigation development came into question. Paralleling the reservation over irrigation investment within the donor community, irrigation was also sliding on the international research agenda. For instance, IWMI's research on irrigation moved to the fringes of the CGIAR's overall research agenda. This was clearly an undesirable trend, especially because it was based on insufficient knowledge not only on the impacts of past investments in irrigation but also on the implications of the new challenges facing the irrigation sector in particular and agriculture sector in general.

In its responsibility as a leading knowledge center on water, food and environment, IWMI has led a number of research initiatives to inform and refine global research on, and investment in, water and agriculture. As noted elsewhere in this volume, IWMI itself has undergone significant transformations over the past 10 years in terms of its research agenda and organizational structure to better reflect the broader water, food and environment nexus as well as the changing realities of the water challenges in general. In this process, knowledge and capacity development initiatives, especially in thematic areas where there are serious gaps, have emerged as the core activity of IWMI. The CA is one of these initiatives.

The CA is a 6-year international research and capacity-building program that takes stock of the costs, benefits and impacts of the past 50 years of water development for agriculture, the water management challenges communities are facing today and solutions people have developed. The results of the assessment will enable farming communities, governments and donors to make better and more realistic investment and management decisions to meet food and environmental security objectives both in the near future and over the next 25 years.

From an organizational perspective, the CA is carried out through the CGIAR System Wide Initiative on Water Management (SWIM) with the involvement of a larger number of researchers spanning across CGIAR institutions, their national partners and others organizations. This chapter will describe the evolving scope and coverage of this research program, outline the conceptual framework, list the research questions for which answers are sought and, finally, provide a synthetic overview of the findings to date on many of the important issues related to water, food and the environment.

8.2 EVOLVING SCOPE AND CONTEXT

The CA essentially evolved from the initial work and research networks created under SWIM. The first phase of this initiative, referred to as SWIM-1, was initiated in 1995 by IWMI and its CGIAR partners. SWIM-1 formed part of the overall paradigm shift for IWMI and corresponded with the Institute's shifting focus from 'water for irrigation within a system' to 'water for agriculture within a basin'. SWIM-1 also took the first critical steps in building a major collaborative program for dealing with the broader issues of water management and agricultural production. In line with the overall objectives of both IWMI and CGIAR, SWIM-1 focused on generating high-quality research and policy works on a range of key topics on agricultural water use, especially from a multiple-use perspective within the basin context. Research issues covered under the SWIM-1 program included: water accounting, salinity management, water-land relations, water productivity (i.e., more rice - less water), multiple uses (i.e., more use-same water), water harvesting and basin-scale modeling. Many of these issues have also eventually entered into IWMI's research agenda.

While the SWIM-1 research paradigm and scope represented a clear shift of water research within IWMI in particular and CGIAR in general, several key issues were still largely outside the program's research ambit. Notably, the second World Water Forum in 2000 brought attention to some of these issues, particularly as they related to the water-agriculture-environment-livelihood interface (see HRH Prince of Orange and Rijsberman 2000). For example, IWMI's contributions to the Forum concentrated on key global issues around water supply and demand, which included an examination of the water implications of demographic transition, dietary changes, trade and other economic policies, land use patterns, efficiency and productivity on irrigated and rain-fed agriculture, groundwater depletion and inter-sectoral competition, especially from urban expansion and environment requirements (IWMI 2000).

The last issue is particularly important, as historically there has been insufficient understanding of the implications of inter-sectoral competition for food and livelihoods. IWMI argued that although policy attention is usually focused on the growing but overall small share of urban water needs, the main area of competition is between water for food and water for environment (Rijsberman 2002; Rijsberman and Molden 2001; Molden and de Fraiture 2004). Water management in agriculture holds the key to find this balance both by lessening the irrigation demand and by reversing water-related environmental degradation. Clearly, the central issues here relate to water productivity, food security, water-based livelihoods and resource sustainability; and these issues have to be addressed by treating irrigated, rain-fed and wetland systems together within a landscape context such as river basins, catchments, or watersheds (Molden *et al.* 2001b; Penning de Vries *et al.* 2002; Molden 2002).

Responding to these new challenges, SWIM was remodeled in 2000 with a broader scope and enhanced research network that went beyond the CGIAR centers to include other national and international partners. Following from this change, the CA (or SWIM-2) has become a program for building a sound knowledge base on water, food and environment. The CA also directly links with two other global programs hosted by IWMI with the aim of refocusing global attention on the water-food-environment nexus: the Dialogue on Water, Food and Environment and the CPWF, both described in chapter 2. Within this multi-program scheme, the CA provides inputs for the CPWF research and funding prioritization as well as benefits from the synergies flowing from both the Dialogue and the CPWF. Similar synergies exist between the CA and IWMI's five thematic areas described in this volume, and thus the program has been an integral part of the Institute's overall research and outreach programs.

The key output of the CA will be a 'State-of-the-World' assessment report (the synthesis volume) and a set of options backed by hundreds of leading water and development professionals and users. Through an extensive process of consultation and review, the Assessment will synthesize the CA research with inputs from other research programs and from a wide range of stakeholders. This process will create a unique fusion of state-of-the-art science, on-the-ground experience and traditional knowledge from communities around the world. The process aims to involve as many stakeholders as possible in order to guarantee the quality and relevance of the Assessment as well as encourage ownership of the Assessment findings.

8.3 CA FRAMEWORK

CA is not a program of new research but rather an assessment and synthesis of what is known already from past research and for the identification and prioritization of additional areas on which we need to know more for future research. Since the subject is vast and diverse, there is an understandable need for a framework to prioritize the issues and to channel the energies for their assessment. The framework for the CA is delineated by a set of key questions that capture some of the most important issues related to water, food and the environment, namely:

(1) How much water will be needed for agriculture?
(2) What are the options and their consequences for improving water productivity in agriculture?
(3) What have been the benefits, costs and impacts of irrigated agricultural development and what conditions those impacts?
(4) What are the consequences of land and water degradation on water productivity and the multiple users of water in catchments?
(5) What is the extent and significance of use of low-quality water in agriculture (saline and wastewater), and what are options for its use?

(6) What are the options for better management of rainwater to support rural livelihoods, food production and land rehabilitation in water-scarce areas?
(7) What are the options and consequences for using groundwater?
(8) How can water be managed to sustain and enhance capture-fisheries and aquaculture systems?
(9) What are the options for integrated water-resources management in basins and catchments?
(10) What policy and institutional frameworks are appropriate under various conditions for managing water to meet goals of food and environmental security?

These questions taken together form the overall framework of the CA. It is important to note that the CA will only be completed in 2006, and thus not all of the research questions have been fully addressed. As a result, this chapter focuses primarily on questions 1, 2, 3 and 10, for which the bulk of the research has been completed.

8.4 DETERMINING AGRICULTURAL WATER NEEDS

The framing of questions for the CA is: How much water will be needed for agriculture to meet global food and livelihood challenges? While this question may seem simplistic on the surface, it forms the very core of the water for food and water for environment debate and how society resolves this question is central in the search for a balance between food and environmental security. Clearly, from the perspective of the poverty and hunger-related Millennium Development Goals, the question goes beyond the usual concern of more water and more food to encompass the broader roles of water in poverty alleviation and livelihood generation. Further, the question of how much more water is needed for agriculture will not have any universal answer but rather will vary depending on the particular basin or country, and the values of people.

As to the global demand for water in general and agriculture in particular, different groups of global modelers have provided estimates based on varying assumptions (Rijsberman 2000). Table 8.1 provides these estimates under the business-as-usual or the base scenario. As can be seen, the results of Shiklomanov 2000, IWMI 2000, FAO 2002 and IFPRI (Rosegrant *et al.* 2002) show an increased level in the need for water in agriculture, while that of Kassel University (Alcamo *et al.* 2000) shows a drop in the amount of withdrawals for irrigation.

Table 8.1 Projected increases in global water withdrawals (km^3).

Source	Total water withdrawals			Total irrigation withdrawals		
	1995	2025	Percent increase (1995-25)	1995	2025	Percent increase (1995-25)
IWMI	3,324	4,368	31	2,469	2,915	18
IFPRI	3,906	4,772	22	1,435[a]	1,492[a]	4
FAO	3,600	-	-	2,128	2,420	14
Kassel	3,572	4,091	15	2,465	2,291	-7
Shiklomanov	3,765	5,137	36	2,488	3,097	24

Note: [a] Total irrigation withdrawal figures of IFPRI are actual consumptive use in irrigation.

Sources: FAO 2002; Rosegrant *et al.* 2002; IWMI 2000; Alcamo *et al.* 2000; Shiklomanov 2000.

The reason for the variance in the estimates essentially reflects the differences in the assumptions and viewpoints among the modelers. For instance, the projection by Shiklomanov (2000) considers present trends and extrapolates them to the future under the business-as-usual scenario that forms an important bound for the problem. The Kassel estimate is based not on food but on environmental considerations, especially on the need to reduce stress on river systems. It also does not take into consideration the important issue of return flows. In fact, projections of Shiklomanov and Kassel form, respectively, the upper and lower bound of the envelope of solutions, with the former failing to incorporate the role of technological changes and the latter failing to account for the water for food and the role of return flows.

In contrast, the estimates of IWMI, FAO and IFPRI are more directly concerned with the water needs of agriculture and also incorporate the effects of productivity improvements. In fact, IWMI's estimate expects an increase in water use efficiency and productivity of irrigated agriculture, but it is somewhat pessimistic about the water savings from rain-fed agriculture[2] (see Table 8.2). Besides, this projection is also based on the assumption that most countries will opt for food self-sufficiency rather than rely on trade. The FAO estimate is slightly more optimistic about the gains in rain-fed areas and hence it suggests a reduced need for irrigation. The IFPRI estimate assumes a very slight increase in consumption (evapotranspiration) from irrigation but large gains in rain-fed

[2] A tricky nuance here is the definition of rain-fed agriculture. If people use small-scale water harvesting or supplemental irrigation, IWMI would consider this a conversion to irrigation. Others would classify this as an upgrading of a rain-fed system. Clearly, there is a continuum between pure rain-fed and pure irrigation systems, and not a sharp divide between the two.

areas. Notably, since much of the gains occur in developed countries, it is expected that trade will expand to distribute food across the globe.

Table 8.2 Global cereal production from rain-fed and irrigated areas: Estimates of baseline values and projected increases.

	Cereal production (million metric tons)						Percent increase		
	1995			2025			1995-25		
	Irrigated[a]	Rain-fed	Total	Irrigated[b]	Rain-fed	Total[c]	Irrigated	Rain-fed	Total
IWMI	727	997	1,724	1,331	1,079	2,410	83	8	40
IFPRI	742	1,033	1,775	1,161	1,453	2,614	56	41	47
FAO	727	997	1,724	1,133	1,369	2,503	56	37	45

Notes: [a] Irrigated cereal production as a percent of total in FAO is taken as 42.15%, consistent with the growth rate of total cereal production of developing countries.

[b] FAO projection of the share of irrigated production of the 2025 total is taken as 45.29%, based on the FAO estimates of share of irrigation to total cereal production in developing countries.

[c] FAO projected a growth in total cereal production at 1.25%. We used this growth rate on the 1995 value of irrigated cereal production, estimated by IWMI, to obtain 2025 values.

Sources: FAO 2002; Rosegrant *et al.* 2002; IWMI 2000.

Adding another dimension to the debate, Rockstrom *et al.* (1999, 2003) argue that the above estimates focus only on the 'blue water' sources (i.e., rivers and aquifers) and neglect an important part of the equation, the water used in situ in rain-fed systems. Rockstrom *et al.* (1999) estimate that the in situ use of rainwater by rain-fed agriculture is about 4,500 km^3 as against the blue water withdrawal of only 2,500 km^3 for irrigated agriculture. This is an important consideration in global water use, as 60% of cereal is presently produced on rain-fed lands. In areas such as the Sub-Saharan Africa, irrigation is of little importance, with 95% of cultivated lands under rain-fed agriculture (FAO-STAT 2000). Even in Asia, with the largest share of global irrigation, 65% of cultivated lands are under rain-fed agriculture. Most of the world's 900 million rural poor are directly dependent on rain-fed agriculture for their food and livelihood needs (IFAD 2001). Managing water in rain-fed systems is of direct concern to the overall water management picture. Indeed how much more irrigation is required depends on the productivity of rain-fed agriculture.

In many ways, the global water supply and demand assessments discussed above are insufficient partly for data limitations and partly for the nature of the assumptions. They also gloss over nuances that make all the difference for people highly dependent on water for food and livelihoods. There is a clear need to explore the issues much deeper than before, especially by breaking them into pieces that can be understood better by society and allow for more realistic policy decisions and investment choices. The CA, when completed, is expected to play precisely this role. While there are clear trade-offs between food and environmental needs of water, there are also complementarities as, for instance, well-maintained wetlands and forests that can, and do, provide important food and livelihood options. At the same time, meeting the growing water needs of food production need not automatically guarantee poverty reduction and nutrition security. Therefore, direct poverty and rural development roles of water development and management have to be explicitly taken into account. Similarly, additional production in rain-fed areas will certainly be important for rural livelihoods and for the water-environment balance. However, it is uncertain how much food and livelihood gains can be expected from rain-fed lands. Improvements in water productivity, in the sense of 'more food with less water', are thus critical to reduce additional diversion both in rain-fed and irrigated systems.

8.5 WATER PRODUCTIVITY: OPTIONS AND EFFECTS

The challenge then is whether it is possible to reduce additional allocations to agriculture to 0% at a global scale over the next 25 years and, thereby, meet the growing water needs of the environment. At the same time, such reduction should not risk the food, poverty alleviation and livelihood roles of agricultural water allocation. This is one of the central questions that the CPWF seeks to answer in basin-specific contexts through the promotion and funding of selected research proposals (Challenge Program on Water and Food Consortium 2002). For the CA, this is equally an issue of top priority, especially since the answer is rooted in a strategy where reductions in agricultural water allocation are the result of a simultaneous improvement in water productivity both in irrigated and rain-fed agriculture. It is generally argued that water savings in agriculture can be an effective supplement, if not an alternative, to a policy of developing new supplies This 'demand-management' approach requires that with the same amount or less water, additional food must be grown to meet increasing food demands. Here lies the strategic and practical significance of the concept of water productivity, popularly known in terms of the slogans as 'more crop per drop' and 'more food with less water'.

A rough estimate shows that if water productivity can be increased, on average, by 40% over the next 25 years, then, it is indeed possible to make additional global diversions to agriculture approach 0%. The issue of balancing the water needs of agriculture and environment, therefore, boils down to the issue of how to achieve improvements in water productivity in agriculture. The first critical step is to establish a common understanding of the concept of water productivity itself. In the broadest sense, the concept of agricultural water productivity considers values derived from water use by forests, livestock, fisheries and crops. Further, the concept allows for cross-sectoral analysis, where values of different uses are considered. The research reported in Kijne *et al.* 2003, among others, makes an important contribution in this respect. Kijne *et al.* cover not only the definitions and concepts but also some important examples of agricultural water productivity under various settings, ranging from pure rain-fed to fully irrigated systems.

8.5.1 Policy options and research needs

A number of technical and policy options and strategies have been proposed to increase water productivity including:

(a) Crop breeding.
(b) Low-cost, precision irrigation technologies.
(c) Supplemental and deficit irrigation and water harvesting for rain-fed areas.
(d) 'Deficit irrigation' by applying less water than the maximum crop demand, but at critical growth periods in irrigated areas.
(e) Improved field-level agronomic practices.
(f) Improved irrigation-management practices.
(g) Reducing land degradation.
(h) Integrating recycling and reuse into basin-management practices.
(i) Reduce non-beneficial depletion[3] by minimizing flows to sinks and non-beneficial evaporation, taking care that these do not serve other beneficial uses.
(j) Integrated natural resources management within basins that enhances productivity of water across uses.
(k) Policies and institutions for the right set of incentives.

Given the linkages among these options, there are a variety of interconnected pathways that can be used to improve the productivity of water. To be successful, however, integrated strategies must be tailored to the needs of specific regions and river basins.

Because of the important potential for improving water productivity, the concept has emerged as an organizing principle for the research programs of both the CA and the CPWF. As described in the CA book by Kijne *et al.* (2003),

[3]Depletion is the use of water that renders it unavailable for further use within the hydrologic cycle (Molden 1997). Depletion is caused by evaporation, transpiration, diverting flows to sinks, pollution and incorporation of water into a product.

further research is required on many aspects of water productivity (see Box 8.1), and many of the related research questions have been adopted by the CPWF. Research results from the CA, however, are already highlighting innovative approaches to increasing water productivity, some examples of which are provided in the next section.

Box 8.1 Where is more research needed on water productivity?

(a) Crop breeding for drought-tolerance, water conservation and ability to thrive on low-quality water.
(b) Understanding the interaction between water-management practices at different levels—field, system and basin.
(c) Co-managing water for agriculture and the environment.
(d) Co-managing surface water and groundwater.
(e) Appropriate technologies and practices for improving water productivity at field and irrigation-system levels.
(f) Policies and incentives needed to implement water-saving technologies and practices.
(g) How to manage irrigation water for multiple uses—for crops, for domestic use and for other income-generating activities.
(h) Tools and models to support responsible decision making for valuing the productivity of water in its various uses and examining trade-offs.

8.5.2 Examples of innovative practices

An important part of the CA is to search for innovative approaches to increase water productivity, especially those that have potential for poverty reduction or that could release water for ecosystems and other needs. We summarize below three such examples, one each from China, Central Asia and Sub-Saharan Africa.

In China, where scarcity and competition for water are intense, research is offering important lessons to decrease the amount of water needed to grow rice while still improving yields. The research-conducted by the International Rice Research Institute (IRRI), IWMI, Australia's Commonwealth Scientific and Industrial Research Organization (CSIRO) and China's Wuhan University-has shown how improving the productivity of water was made possible through on-farm water-saving irrigation practices, ample water recycling, pricing water and strong institutions (Barker *et al.* 2001; Loeve *et al.* 2004; Dong *et al.* 2004). In the Zhang He Irrigation System, where the research took place, rice grown per unit of water diverted from the main reservoir tripled through a combination of yield increases and reduced water applications. This has enabled water managers to shift water out of agriculture to meet growing municipal and industrial demands.

With significant investment in water harvesting, conservation tillage and supplemental irrigation during short dry spells, yields of staple food crops could be more than doubled in many areas of Sub-Saharan Africa (Rockstrom *et al.* 2003). Low-cost drip kits increase the value of agriculture per unit of water by providing a means for farmers to grow vegetables with a limited source of water (Namara *et al.* 2005). ICARDA studies in Syria have shown that applying 50% of the supplemental irrigation requirement only reduces yields by 15% (Oweis and Hachum 2003). While these and other cases of innovations for water productivity improvements are interesting, they also raise an equally interesting question about their adoption and spread in other areas and contexts (Namara *et al.* 2003, 2005). In particular is the question of why and how some practices are adopted while others are not. In order to provide answers to this question, it is important to look into the factors and processes that have influenced the adoption decisions. For this purpose the CA is drawing from the IWMI Bright Spots[4] analysis discussed in chapter 4.

8.6 IRRIGATED AGRICULTURE: COSTS AND BENEFITS

Of total agricultural water use, water used for irrigation is the most contentious not only because of its magnitude but also because of the debate as to whether its actual benefits in terms of food production, poverty reduction and economic development can outweigh its social, financial and environmental costs. While irrigation has played, and will continue to play, a central role in providing food security and supporting livelihoods, several questions still remain unanswered. For example, does irrigation remain the pathway to ensure food and environmental security, or does it accentuate the divide between the rich and the poor and exacerbate the process of resource and environmental degradation? The overall picture of the impacts of irrigation is not at all clear. There is ample information on the primary or direct impacts of irrigation in terms of production and income increases. However, there is a lack of empirical evidence about other secondary or indirect impacts on poverty, resource and ecosystem health, and society as a whole. A better and more complete understanding of irrigation impacts will help in making better decisions about how much more irrigation should be developed, particularly in areas such as Sub-Saharan Africa, and how to balance its food and environmental impacts.

---

[4] A "bright spot" is defined as a community or group of individuals achieving higher food and environmental security, through improvements (among others) in their land and water management practices.

8.6.1 Impacts on economics development

Setting the stage, Barker and Molle (2004) traced the evolution of irrigation in South and Southeast Asia identifying three separate geopolitical time periods: the Colonial Era from 1850 to 1950, the Cold War Era from 1950 to 1989 and the new Era of Globalization from 1990 onward. The development of irrigation, whether by colonial administrations or more recently by national governments and lending agencies, has been pursued with a fairly common set of objectives-poverty alleviation, food security, employment, livelihood, increased revenues and growth in value of agricultural outputs. Against this background, the rapid development of irrigated agriculture has helped foster extraordinary growth and change in the rural economies of Asia. But the success of these endeavors has also brought new problems.

The intensification of irrigated agriculture has led to an increase in pollution and environmental degradation. Food grain prices have plummeted with the result that the benefits of irrigation have gone largely to consumers. Farm households have looked to other sources of income from both farm and non-farm sources. As we enter the new era of globalization, farmers and systems operators have adjusted to the challenges of growing water scarcity by exploiting groundwater, recycling from drains and canals, changing cropping patterns and adjusting the timing of water releases. Tube wells and pumps have also become commonplace, giving producers greater flexibility to obtain water when needed. While coping mechanisms such as these may lessen the local impacts of water scarcity, there has been a serious lag in the development of appropriate institutions to deal with the effects of these new realities and adaptations at broader scales.

Has irrigation led to economic development? Bhattarai *et al.* (2003) in a detailed quantitative analysis have shown that improved access to irrigation and rural education has contributed largely to recent productivity and rural income growth in India. Total impacts of irrigation (direct and indirect) to the regional economy are much larger than the farmers' share of benefits in terms of increased land productivity. For example, only 32% of the total economy-wide benefits are actually realized by the typical Indian farmer, whereas the rest of the benefits of irrigation readily percolate into the society as a whole.[5]

Both benefits and costs of irrigation extend far beyond irrigation system boundaries. The multifunctionality concept recognizes the multiple-output nature of agricultural production in which many commodity as well as non-commodity outputs are jointly produced. Examples of beneficial multifunctional uses of irrigation water are groundwater recharge, creation of wetlands and

---

[5] This is equivalent to a multiplier of 3.2, calculated as the ratio of total to direct benefits.

flood protection. Research focused on multifunctionality of irrigation, particularly paddy rice cultivation, has explored what these externalities have been and how to value them. Boisvert *et al.* (2003) and Matsuno *et al.* (2002) conclude that return flows from rice cultivation generate important positive as well as negative externalities, and provide examples of how to value these externalities.

8.6.2 Impact on ecosystems

There is no doubt that resource use by irrigated agriculture has led to ecosystem degradation. Many river systems are drying up near their mouths, the Colorado, Yellow River, Amu Darya to name a few, mainly because of water diverted and depleted by irrigated agriculture. Several studies have focused on the negative impacts of irrigation, particularly on wetland ecosystems. The widely quoted study of Barbier and Thompson (1998) shows that wetlands generated a value per cubic meter[6] for uses such as pastoralism, fishing and recession agriculture, much higher than the $0.04 per cubic meter for irrigation. While the extremely high value for natural wetlands may raise some doubts about the study, the point is still that alternative uses of water should be considered, and that the way irrigation is developed can result in negative economic impacts. Lemly *et al.* (2000) in a global study of wetlands sums it up by stating, "The conflict between irrigated agriculture and wildlife conservation has reached a critical point at a global scale. Many key wetlands are now a mere shadow of what they once were in terms of biodiversity and wildlife production".

River systems have borne the brunt of irrigation development. Hydraulic structures have fragmented rivers, altered hydrologic regimes and the resulting use of water has led to pollution. All this has impacted rivers and the ecosystem services they provide. A global picture of the environmental condition of river systems was carried out by WRI, IWMI, University of Kassel, and IUCN (Smakhtin *et al.* 2004) that developed rules applicable at global scale to understand environmental flow requirements (EFR). The EFR required to maintain a fair health of freshwater ecosystems at the global level is in the range of 20-50% of the mean annual river flow in a basin. It is shown that even at estimated modest levels of EFR, parts of the world are already, or soon will be, classified as environmentally water-scarce or environmentally water-stressed. The total population living in basins where modest EFR levels are already in conflict with current water use is over 1.4 billion, and this number is growing. The necessity of further research in this field is advocated and the directions for

---

[6] They found a value of $12 per cubic meter, but the authors question whether this is a reasonable value.

such research are discussed. The CA is supporting further research locally to understand environmental flows and how they are impacted by irrigation.

Irrigation does not always lead to negative environmental externalities. Bambaradiniya and Amerasinghe (2003) state that traditional rice fields that have been cultivated over a long period of time may be considered as climax communities-sustainable ecosystems that support biodiversity. The picture has changed with modern technologies, including the use of chemicals, optimum water and crop management practices and machinery that have been successful in increasing yields. These developments have caused profound modifications to traditional rice-growing environments. The surveys on biodiversity associated with traditional rice field agro-ecosystems conducted to date have clearly shown that this man-made ecosystem contributes to sustaining a rich biodiversity, including unique as well as threatened species, as well as enhancing biodiversity in urban and suburban areas. But modern practices have imperilled the potential of rice systems to maintain high levels of biodiversity. Finding ways to grow rice and meet economic and poverty alleviation objectives, yet maintaining biodiversity will require the integrated effort of agroecologists, water managers and conservation biologists working with irrigated farming communities.

In a literature review on irrigation and wetlands in developing countries, the Stockholm Environment Institute (SEI) and the University of Peradeniya (Galbraith *et al.* 2005) confirmed that irrigation or activities associated with irrigation can, and do, cause adverse effects to wetland ecosystems. But, they can also result in the creation or enhancement of important wetland ecological resources. Fortunately, under certain circumstances, wetland ecological resources and irrigated agriculture may coexist. Confounding effects of natural phenomena such as droughts and other anthropogenic stressors such as urbanization complicate the evaluation of irrigation impacts. Unfortunately, the measurement of impacts or benefits due to irrigation project implementation ends with the project installation, leaving a gap in our knowledge. There is hope that irrigation can fulfil its food production function and better support or cause less damage to ecosystems. However, we need more information on how irrigation impacts wetland ecosystems and what are the strategies to manage water to meet food production and other ecosystem sustenance objectives.

The CA studies to date provide a mixed but evolving picture. Yes, irrigation has been an important instrument in economic development and poverty alleviation, but the degree of success has varied. Irrigation has undoubtedly negatively affected important ecosystem functions. However, there are notable exceptions, including examples from multifunctionality of paddy-field cultivation, which may provide clues on how to co-manage water to meet food and environmental security needs. It is likely that studies, which start from the 'benefits of irrigation' point of view will tend to overestimate the potential benefits, while others which have the view of 'irrigation-induced degradation' will tend to overestimate secondary costs and underestimate benefits. There is a need for a new, more objective methodological approach, but in many instances data may often be lacking for the successful application of this approach.

8.7 FOOD SECURITY: THE ROLE OF VIRTUAL WATER

The CA explores the issue of identifying appropriate policy and institutional frameworks for managing water to meet food and environmental security goals at various levels, including managing water within irrigation systems and assessing the water and agricultural related impact of national-level trade policies and subsidies. A key tool for national and global analysis is the WaterSIM model developed by IWMI and IFPRI (de Fraiture *et al.* 2003). This builds on the work of IWMI's Podium (IWMI 2000), and the IFPRI impact model (Rosegrant *et al.* 2002) reported above. The model includes food and water use accounting, economics and environmental flow considerations to allow exploration of future scenarios including more or less irrigation and the role of subsidies, prices and trade in agricultural water use. One aspect of this research explores the role of trade and virtual water, which is discussed below.

Virtual refers to the amount of water used to produce agricultural commodities (Allan 1998). As a rule of thumb, a grain crop evapotranspires between about 0.7 to 2 m^3 of water in order to produce 1 kilogram of grain depending on water productivity. Thus importing 1 kilogram of grain is approximately equivalent to importing 1 m^3 of water. Virtual water flows have relevance to water stress, water scarcity and food security, as they reduce the need to use water for food production in importing countries and increase water use in exporting countries, and thus the topic has raised interest in recent global discussions (Hoekstra and Hung 2003). At present, cereals comprise the majority of trade in agricultural products, and so tracking trade in cereals is a good indicator of overall virtual water flows. A strategy suggested is to develop virtual flows of water in water-stressed countries instead of developing new supplies. Figure 8.1 indicates virtual water flows in terms of cereal trade for the years 1995 and 2025.

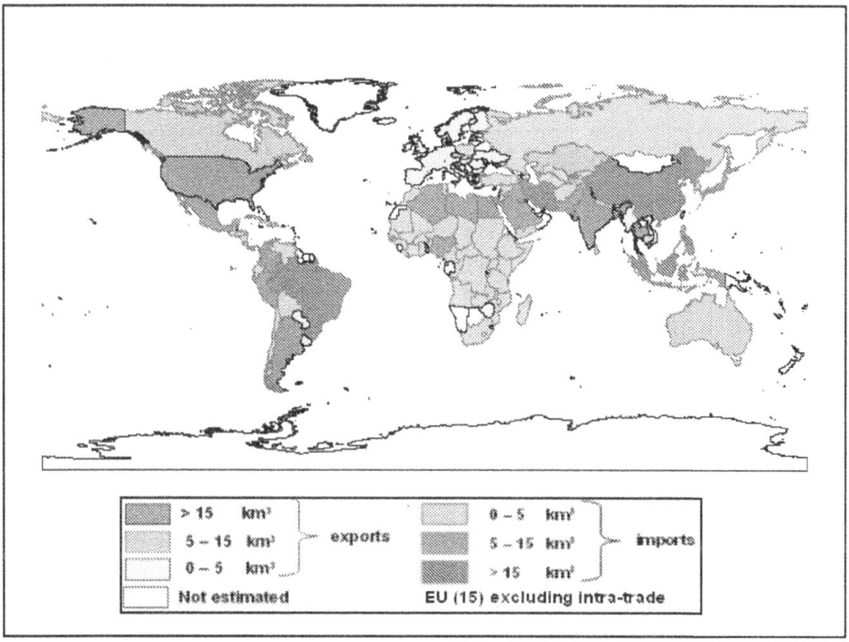

Source: de Fraiture *et al.* 2004.

Figure 8.1 Global cereal trade and virtual water flows, 1995.

What is most striking in Figure 8.1 is the case of China. Although more or less self-sufficient in cereal grain production (only 4% cereal production deficit), China imported a significant amount of virtual water (16 km^3) in cereals in 1995. This is because of the huge population and high consumption levels (2766 cal/pc/day). As a result, small changes in China can shift the global water and food equation. In the years following 1995, China has come closer to food self-sufficiency.

Another interesting case is Japan. While Japan is not water-scarce, the country imports a substantial amount of its cereal requirements (76% of the total consumption) and hence it is a big virtual water importer (25 km^3). Japan is not at all water-scarce, but rather uses trade in food as part of their economic strategy. Most northern African and West Asian countries are physically water-scarce (IWMI 2000) but have high daily calorie supplies due to substantial food imports. Thus virtual water flows into these countries are large and are driven by water scarcity.

Finally, many countries in Sub-Saharan Africa import small amounts of virtual water, even though malnutrition is prevalent and there is no physical water scarcity. In these countries the stakes are high for food and environmental security, but trade and virtual water flows are not significant. One reason is the reliance on subsistence food production, and lack of funds to purchase food by many poor farmers. Additionally, due to the high cost of domestic production, African farmers cannot compete with free-trade imports. Increasing water productivity should help to reduce domestic cost of production[7] and enhance food and environmental security.

De Fraiture *et al.* (2004) postulate that trade could save water globally - more food could be produced with less water - based on the water productivity difference between importing and exporting countries. With only cereals considered, a present savings of evapotranspiration flows of 156 km^3 are realized because of virtual water trade, and under the IFPRI base scenario, an amount of 276 km^3 would be saved in 2025. The reason for trade is not always driven by water scarcity as shown by the large virtual water imports of Japan and South Korea, but may be for other economic reasons.

8.8 CONCLUSIONS

SWIM and the CA work have been instrumental in focusing the agricultural research community on water questions of global importance, and bringing that community to the global discourse on water issues. SWIM was highly successful in producing high-quality outputs on key issues of integrated water resources management. The CA raised the stakes by expanding the scope of SWIM, and focusing it on issues of global concern as reflected in the Millennium Development Goals on hunger, poverty and environment-how water is used in agriculture, and how it can be best used to reduce poverty and improve the environment. This chapter has only touched on some of the initial results of the CA. The four issue areas described above plus the research related to the remaining six key questions of the CA will be presented in much greater detail in a range of products and dissemination fora in 2006.

Upon its completion, the CA will have produced a wealth of materials for various audiences. For water professionals, researchers and students, it will provide a (a) state-of-the-world assessment volume, summarizing key results of the CA research, (b) peer-reviewed research report series, highlighting collaborative work and presenting overviews and issues of policy relevance, and (c) book series presenting the results on various CA questions. For policymakers and investors, it will provide (a) an overview of the Assessment volume, highlighting key research-backed messages, and (b) a set of policy and issue

---

[7] This requires that the costs of technology and management are offset by productivity and production gains.

briefs based on CA research results. For IWMI, the CA provides a strategic assessment of key research areas of focus by providing a participatory- and research-backed scoping mechanism. Upon its completion, the Assessment and its related products will highlight key research needs for the research community involved in water and agriculture. For policymakers, the Assessment results will guide investments in water and agriculture; and for implementers, investors and policymakers, the Assessment will be a milestone from which to measure achievements, and a guidepost providing important directions for action.

9

A DECADE OF WATER RESEARCH AT IWMI: INSIGHTS AND IMPACTS

R. Maria Saleth and Meredith A. Giordano

9.1 INTRODUCTION

In the history of IWMI, the years during 1996-05 form an important period of rethinking, expansion and change. It is also a period of major adjustments, especially with a broader research mandate, a new research paradigm and a new thematic structure. These changes were necessitated in part by the changing character of the water challenge as well as by developments in the global water sector itself, including declining trends in global investment in water development and reduced emphasis on water issues within the international agricultural research system. The changing nature of the water challenge means that the food, poverty and livelihood implications of water go far beyond irrigation management and are fundamentally linked with the broader issues of resource management, including land and environment. From the perspective of IWMI, declining investments in water development and research suggested a

© 2006 IWMI. *More Crop per Drop*. Edited by M.A. Giordano, F.R. Rijsberman, R. Maria Saleth. ISBN: 9781843391128. Published by IWA Publishing, London, UK.

need to reframe the water question to fit with the changing socioeconomic and environmental context. It is this diagnosis that led to a shift in research focus from 'irrigation' to 'water' management in 1996 and, subsequently, broadened its mandate further in 2001 to cover also land management issues functionally linked to water resources.

With the change in the research focus and mandate, the water issue was also framed in a larger context defined by the 'water-food-environment nexus'. This context has enabled research to consider both the trade-offs and synergies evident among various roles of water within a common and consistent analytical framework. Water resources are considered not only in conjunction with land and environmental systems but also in the entire spectrum of its uses ranging from irrigated and rain-fed systems to riverine and wetland ecosystems. The research paradigm centered on 'more crop per drop' was developed to focus on water reallocation and recycling as the main means for increasing water productivity. The paradigm has been further refined over time to generalize its application beyond irrigation and agriculture to cover related socioeconomic, health and environmental benefits. The concept of water productivity is now considered both as an analytical tool to bring complex issues within a simple and tractable framework and as a solution algorithm for designing practical policies to solve allocation issues.

What are the implications of these changes for water research at IWMI? How did the research carried out during 1996-05 respond to the challenges and opportunities presented by these changes? To what extent have these changes enabled IWMI to meet its ultimate missions and strategic goals in the unfolding realities of the global water sector? The previous chapters have attempted to answer these and related questions with an overview of the new research context, critical review of the underlying research paradigm and synthesis of research conducted under various thematic areas. Admittedly, this is not an easy task, particularly given the limited space and the broad research canvas and time coverage. Within these constraints, each chapter succeeded in presenting the relevant results as briefly as possible, yet with enough details and leads that will be useful to researchers, policymakers and donors. This chapter makes an attempt to highlight some of the key insights and also to indicate the impacts of IWMI's works on water research, policy debates and actual practice both at the local and global levels.

9.2 FRAMING THE WATER QUESTION

The nature of water challenge is changing fast with many misconceptions becoming clarified and new realities being unraveled with new knowledge and information. The past tendency to generalize water crisis is giving way to dissect the issue to locate the areas that are really at risk and those where the risk is less to do with physical scarcity but more to do with financial, institutional and political problems. Similarly, it is also clear now that the food and livelihood implications of water scarcity have less to do with water scarcity per se but more to do with the issues of allocation, access and productivity of water. It is true that water needed for food production and rural livelihoods is constantly under threat from increasing diversions to urban and industrial sectors. But, the negative effects on food and livelihoods are due not to this reallocation per se but rather to the inability to increase the productivity of the water, especially in the agriculture sector, which claims the largest share of water at present. Once the central role of water productivity is recognized as the main means for mediating inter-sectoral allocations, the water crisis presents itself more as an economic and technical issue than as a physical problem. Since water productivity links water research with the biophysical, socioeconomic, and technological research, it enables us to recast a water crisis, not as an insurmountable physical outcome but as a solvable question of management and policy.

The environmental dimension, which was somewhat overshadowed in the past by the predominant concern over the food and livelihood implications of water scarcity, has now also come to the forefront. Although the environmental issues were usually considered in terms of the ecological effects of water development, water pollution and land use changes, attention is now on the need for environmental allocations and their socioeconomic implications. Contrary to common perception, environmental allocations are not a luxury. It is indeed an indispensable socioeconomic and ecological necessity for all countries. Given the magnitude and significance of water-based food and livelihoods that are being lost due to the degradation and disappearance of floodplain, riverine and wetland ecosystems in these countries, the socioeconomic implications of the environmental water crisis can be as large as, if not larger than, its ecological consequences.

There are important trade-offs between the water demand for food and that for environment in particular contexts, but they need not necessarily lead to irresolvable conflicts, as being often presented in popular discourses on the subject. It is, in fact, the unfounded and somewhat magnified fear of food shortage that supports the current pattern of unsustainable water use in agriculture, which does not allow us to realize the huge social benefits that are possible with a reallocation of water away from agriculture. Improvements in water productivity can enable us to produce more outputs and values with less water, which can, in turn, make it possible to release enough water for meeting environmental sustainability requirements. It is on this premise that IWMI frames the issues within the context of the 'water-food-environment nexus'.

9.3 REFINING THE RESEARCH PARADIGM

'More crop per drop' has been the central paradigm underlying IWMI research since 1996. The paradigm is defined by three components, i.e., basin focus, recycling and water productivity. While the first two components define the analytical and conceptual context, the last component constitutes the central mechanism for guiding water allocation and use pattern. Reliance on the hydrological unit as an analytical context enables us to present the ideas of 'open' and 'closed' basins that correspond, respectively, to cases where there is scope for additional water development and where there is no such scope. In the challenging cases of closed basins, new uses and users can be accommodated mostly with reallocation from others, as reuse options, though possible, are neither costless nor infinite.

While reuse options such as wastewater use and water recycling are important, they cannot alone be sufficient to meet the massive need for water reallocation within and across basins. Under this condition, our ability to release water of the magnitude required for meeting new uses and users depends ultimately on effecting productivity improvements in existing water uses. Thus, substantial improvements in water productivity, especially in water-intensive sectors, such as agriculture, have the potential to open up even 'closed' basins. It is in this sense that water productivity, as the third component of the research paradigm, assumes its critical significance.

While the 'more crop per drop' paradigm continues to govern IWMI research, there have been some major refinements, especially on the concept and role of water productivity. The idea of 'more crop per drop' is often interpreted literally to mean only 'crop', particularly in irrigated agriculture. But, as a metaphor for water productivity, it covers the per unit benefits of water in all sectors and contexts. Conceptually, it covers all water-related benefits - both direct and indirect - in irrigated, rain-fed, reuse, and riverine and wetland contexts. These benefits cover not only the food, income and livelihood gains but also the health and environmental impacts. In short, the concept should be applicable within the whole spectrum of the hydrological cycle. In this sense, water productivity is treated as a powerful analytical tool to relate water uses in different uses, sectors and regions within a common framework. Since water productivity is sensitive to technical interventions and economic policies and since changes in water productivity can be measured fairly accurately, it is also considered as a solution algorithm. Thus, productivity variations among technologies, institutions, policies and practices can be used as a basis for making decisions on the appropriate mix of interventions suitable for different contexts. In the refined research paradigm, therefore, water productivity assumes a larger conceptual and analytical role and significance than was the case before.

The refined research paradigm, though elevates water productivity as the guiding principle for analysis and decisions, also recognizes, at the same time, that water productivity is not an end in itself but a means to tackle the central issues of water reallocation across sectors. The focus on water productivity does not compromise the importance of equity and distributional issues. It is for this reason that water productivity mapping proceeds along with poverty mapping, where the issues of water access and impacts on income and livelihoods are taken into account. The refined research paradigm is also resilient and sensitive enough to accommodate both the sustainability concerns - related not just to water but also to land and other resources - as well as the distributional and equity aspects. Thus, technical interventions and policy options are evaluated by simultaneously considering their impacts on water productivity, poverty and livelihoods, health, and resource and environmental sustainability. In this way, the efficiency and productivity objective is balanced with the equity and sustainability concerns. It is this approach that forms the conceptual basis, which is clearly reflected in the new research agenda and thematic structure.

9.4 THEMATIC RESEARCH: OVERVIEW AND INSIGHTS

The adoption of a broader research mandate and the recognition of the larger roles of water mean a vast research canvas. The contextualization of the water question within the water-food-environment nexus and the conceptualization of water productivity as an organizing framework do delineate the general contour of the research agenda. But, still, there is need to prioritize the research issues to invest the available research efforts and resources in a more effective and organized manner. To achieve this, IWMI relies on Strategic Plans and Medium-Term Plans that help to target, organize and monitor research during a given period. The Strategic Plan 2000-05, for instance, identified five thematic areas: (1) Integrated Water Resources Management, (2) Smallholder Land and Water Management, (3) Sustainable Groundwater Management, (4) Water Resourccs Institutions and Policies, and (5) Water, Health and Environment. These themes were formed in 2000 with the reorganization and reorientation of the research programs as part of the 2000-05 Strategic Plan. Added to these five themes is also the CA, a time-bound and multi-organizational research initiative led by IWMI. The main highlights and insights of research under the five themes and the CA are given below.

9.4.1 Integrated water resources management

Research under this theme is based on an integrated approach, where irrigation, though a dominant use, is only one of several equally important uses of water, while the water resource itself is viewed as part of the larger biophysical resource system. This theme is, in fact, a natural corollary of the broader research mandate and the new research paradigm centered on water productivity, water allocation and environmental concerns on a spatial scale. A major part of the research under this theme has been devoted to the development of tools and methodologies needed for the practical application of the integrated approach to water resources management at the basin level. A range of water and land productivity indicators were developed and empirically applied to compare water productivity across regions, sectors and uses.

Since water productivity is evaluated at the basin scale, the concept was generalized to cover all water uses within the entire spectrum of the hydrological cycle and all benefits, ranging from crop yields both in irrigated and rain-fed systems to fish outputs, health benefits and ecological contributions. A series of water productivity case studies were conducted in basins located in countries as different as Sri Lanka, India, Pakistan, China, Turkey, Iran and Uzbekistan. Notably, water productivity became an integral part of any research - whether hydrological, economic, institutional, or even, gender analysis - at IWMI and its levels and impacts are used as criteria for selecting policy options and practical strategies.

Methodological tools such as water accounting and hydronomic zoning were also developed and applied to assess the demand-supply profile and use composition of basin water resources. Besides their roles in facilitating the evaluation of water productivity, these tools were also used to generate and evaluate scenarios for spatial and sectoral water allocations within basins. Case studies conducted in Sri Lanka, Pakistan and India have demonstrated how water accounting can be used to estimate both the non-beneficial water withdrawals as well as the potential incremental benefits that can be realized from their reallocation across subbasins and sectors. The method of hydronomic zoning introduced a spatial dimension to the assessment of water resources and the identification of water management strategies. It enables us to develop location-specific strategies that can better tackle the intra-basin variations in hydrological and agronomic conditions. In view of such important roles, the water accounting framework and hydronomic zoning approach are used as an integral part of basin-scale modeling studies conducted by IWMI in different locations around the world.

While the integrated approach is conceptually appealing and practically relevant, translating it in the empirical context, even in a few basins, is not an easy task. Research in this area relies heavily on modeling studies, which demand a considerable amount of diverse information. But, such information is not usually available to the needed extent in many developing countries. Considering the time and resource requirements, IWMI has identified select basins, known as the 'benchmark' basins to concentrate its research efforts and to consistently build on the information base. To gain from past modeling experiences, IWMI is also collaborating with other organizations (e.g., Stockholm Environment Institute and International Food Policy Research Institute) in building models and refining their modules to suit the conditions of the focal basins.

Although it is still in an evolving state, IWMI has already made a beginning in building an electronically retrievable spatial database for a number of basins by combining data from secondary sources, remote sensing and simulation exercises. One critical area where IWMI research has made a major contribution relates to the linking of conventional nn with those from simulation and remote sensing. The modeling studies conducted in the Gediz Basin in Turkey and Zayandeh Rud Basin in Central Iran, for instance, have demonstrated how such information can be used for developing water allocation and use scenarios. This approach is also followed in the currently ongoing modeling studies in the Krishna Basin in India and the Olifants Basin in South Africa.

9.4.2 Smallholder land and water management

Although research on smallholder issues has a long history at IWMI, it became a formal theme only since the adoption of the broader mandate and especially since the organizational merger of the International Board for Soil Research and Management with IWMI in 2001. Research under this theme has tried to promote a holistic understanding on water, land and natural resource management issues in the particular context of smallholder agriculture. It has also helped to analytically link irrigated systems with rain-fed agriculture practiced under varying conditions of informal irrigation and in situ water use for crop and livestock production.

A characteristic feature of smallholder agriculture in Asia and Africa is its predominant concentration in upland, rain-fed and other fragile regions facing serious production risks and ecological consequences. The objective of improving land and water productivity in these contexts is, therefore, linked closely with soil and water conservation as well as smallholders' access to land and water resources, agricultural inputs and commodity markets. The issue of access is particularly important, as smallholders are both the late entrants into the resource markets and also the laggard groups yet to reap the real benefits of the Green Revolution. Research under this theme, therefore, has a dominant focus on the approach and options for improving access, conservation and productivity of land and water resources among smallholders in rain-fed and fragile regions.

While smallholder agriculture is beset with a number of economic and resource-related problems, it also presents major opportunities, especially for linking poverty alleviation with food security and resource conservation both at the farm and landscape levels. Besides their immediate effects on income and livelihoods, on-farm resource use efficiency and conservation also have the long-term role of reducing catchment and downstream consequences such as soil erosion, sedimentation, land loss, water pollution and degradations of aquatic ecosystems. Research work, done both within and outside IWMI, confirms, with numerous instances from Asia and Africa, that these opportunities are real, substantial and achievable.

Based on a number of large-scale research projects and specific case studies in South and Southeast Asia and southern Africa, IWMI has developed the methods and identified the technologies for exploiting these opportunities. These include the methods for land and water conservation on sloping lands, fertility improvement in acidic soils, and rejuvenation of degraded soils, and the technologies for providing water access to small farms and home gardens such as the treadle pumps and miniature pumps, small-scale multiples use systems, homestead rainwater harvesting, and wastewater reuse. The empirical results of these methods and technologies, in terms of their productivity, livelihood and conservation effects, are quite remarkable as are the prospects for up-scaling them on a landscape and catchment scales.

The major question, then, is how to scale up their adoption on a larger scale. The economic gains of land and water conservation methods and technologies can themselves be central in their adoption and spread, but more proactive approaches are still needed for their up-scaling. This clearly requires building networks and creating partnership among researchers, training and extension organizations, and community members. The experiences with the network approach show that the research-training-extension network itself can provide a powerful framework for promoting land- and water-conservation methods and technologies. The network of international research organizations, national research and extension systems, NGOs and local communities helps to minimize research costs, strengthen the research-training-extension linkages and overcome the limitations from the fragmented nature of location-specific research. The success of the network research projects on soil and water management issues in Southeast Asia and Africa attests to the importance of networks as a means to improve both the effectiveness of research as well as its impacts on smallholder communities.

Often, the dire consequences of resource degradations themselves prompt many small farm communities to adjust past, and adopt, new resource use practices. There are many such instances in upland and rain-fed regions throughout Asia and Africa. To document and study these locally evolved innovative practices, IWMI has developed the idea of 'Bright Spots' as an analytical framework. What distinguishes the 'Bright Spots' approach from that based on 'best practices' and 'success stories' is the fact that attention is not just on the best practices but equally also on their socioeconomic, biophysical and institutional context. This context is very important, as it enables us to identify the drivers of change, which can provide an understanding of the factors that can be used as policy instruments for up-scaling innovations. A variety of 'Bright Spot' cases compiled by IWMI has the common message that yield improvements are possible across a wide range of farming systems. Since the major gains are indicated in areas where the gap between potential and actual production is the greatest, low productivity systems are not to be dismissed as hopeless cases but to be viewed as the bright spots of opportunity.

9.4.3 Sustainable groundwater management

Scientific knowledge on groundwater is somewhat disproportionate to the growing economic importance and livelihood significance of this resource in most arid parts of the world. This is particularly evident in the continuing asymmetry between the knowledge on the hydrogeology of the resource and the same on its economic, social and institutional dimensions. With the widespread depletion of this resource, this knowledge gap can have serious effects on the millions of rural families relying heavily on groundwater for their income, livelihood and survival. The challenge now is to identify and promote economic policies and institutional changes, which are necessary to sustain the massive welfare and equity gains from groundwater irrigation while minimizing the social costs of intensive and inequitable exploitation of the resource. It is this challenge that set the overall thrust of IWMI research on groundwater in general as well as the IWMI-Tata Water Policy Program in particular. Although groundwater research is relatively new for IWMI, starting only with the Strategic Plan of 2000-05, it has made considerable advances in terms of issue coverage, methodological framework and policy development.

The framework of groundwater socio-ecology - which captures the evolution, growth and decline of groundwater dependent farming and socioeconomic systems - was used to trace the linkages among the levels of groundwater development, agricultural growth and socioeconomic development. This framework is central for creating the 'big picture', where the relative income and livelihood roles as well as the differential institutional and environmental contexts of groundwater are demonstrated and contrasted with that of large-scale surface irrigation systems. The application of, and the insights from, this framework, though relevant for most groundwater-based economies, are particularly appealing for Asia, where the poverty and livelihood linkages of groundwater are stronger and more vivid.

In this region with a dominance of innumerable small farms, even a small disturbance in the groundwater system can cause misery to millions of people. By the same token, in water-short areas, even a marginal increase in the access to groundwater can make a major difference on the food and income levels of the poor. This is clearly evident in several of the studies conducted on groundwater access in the Gangetic Basin, particularly in Bihar and the eastern Uttar Pradesh, India. Much more dramatic, however, is the role and impacts of the manually operated treadle pumps in India and Bangladesh and the miniature pumps in South Africa. For thousands of poor families with tiny plots, access to groundwater facilitated by these small and affordable technologies represents a switching point between poverty and prosperity.

Besides technologies, appropriate institutional arrangements for groundwater governance are also critical both to control unsustainable depletion and improve equity in resource access and use. Research, both within and outside IWMI, has investigated the effects of direct regulations such as those related to the location, design and depth of wells, but has concluded that they cannot be effective without appropriate and user-oriented institutional arrangements needed for local enforcement and monitoring. Although the application of economic instruments such as pricing will have practical difficulties for groundwater regulation, a surrogate approach based on energy pricing is more likely to succeed in view of the energy-irrigation nexus. However, the most promising avenue, which is already being explored in many contexts, even, by small farmers, is application of water-saving technologies and resource-conserving practices, including supplemental irrigation, conjunctive use, crop pattern changes and water harvesting.

The key driving force behind most coping strategies is the idea of water productivity itself. Resource scarcity has informally inculcated this among users, which, in turn, has created necessary economic incentives and endogenous pressures for exploring unconventional approaches. In this sense, the emergence of groundwater and pump-set rental markets in India, Pakistan and Bangladesh can be seen as one of these coping responses. The success and positive benefits of these markets as well as the turned-over tube wells in Gujarat, India underline also the larger issues of involving users in groundwater regulation and management. Yet, for promoting and up-scaling all these local strategies, macro institutional arrangement are still necessary, as demonstrated by the role of government policy reforms in promoting efficient groundwater use in Bihar and the groundwater rights reforms on farming system choice in the North China Plains. The research based on the comparative analysis of groundwater institutions and governance in South Asia, North China, Mexico and Africa has tried to identify the core set of macro institutional and policy conditions, which are essential complements and promote local-level institutional changes and coping strategies.

9.4.4 Water resources institutions and policies

There is international consensus now that the global water crisis is less of a crisis due to physical limits but more of a crisis due to faulty institutions and governance arrangements. In most developing countries, water institutions, as defined by water laws, water policies and water organizations, are too dated to respond to the changing nature of water challenge. Since institutional and policy issues assume critical importance in almost all facets of water, land and environmental management, they deserve a more independent and focused attention in their own right. Issues related to their nature, evolution, appropriateness and sources of change are important to understand their present roles and future changes. Given the critical role of institutions and policies in the impact pathways and delivery channels of water-related development interventions, their analysis is indispensable to understand the food, poverty and livelihood implications. In this respect, the analysis of the impacts of more-specific individual, legal, policy and organizational components such as water rights, water pricing, irrigation investment and management decentralization is as important as the general impacts and performance of water institutions using cross-country comparison and applying institutional and political economy theories.

IWMI research on water institutions and policies, over the years, has covered both their micro and macro dimensions with a significant level of success in terms of both methodology developments and empirical applications. It has evolved from an almost exclusive focus on the institutional requirement for irrigation management to cover now a wide range of policy and institutional issues relating to the development and management of water resources at the national, basin and global levels. Low cost recovery, poor maintenance, and physical and performance deterioration used to be, and still are, the dominant issues in most state-managed large irrigation schemes in Asia and Africa. Since organizational decentralization and management transfer can be powerful institutional means to address most of these problems, a major part of international research, both at IWMI and elsewhere, was devoted to the analysis of these institutional options both as a solution to the performance crisis and as a means for empowering farmers.

Research has indicated that irrigation management transfer, though essential for performance improvement and rural empowerment, can be more effective only with concurrent changes in related spheres, in particular, the legal, policy and organizational aspects at the macro levels. The work on the design of basin institutions and cross-country analysis of water institutions can help to provide a more complete understanding of the interaction and dynamics of local, basin and national institutions. Similarly, the methodological research on the structure and environment of water institutions and the empirical analysis of water institutional reforms across countries are useful to understand the process and sources of change and also to develop reform design and implementation principles.

Some of the changes needed to create and strengthen micro-macro linkages within water management are related to specific policies while others require sector-wide reforms. Water pricing reforms, for instance, are important not only for cost recovery but also for providing justification for new investments and facilitating resource allocation across uses and users. But, for them to be effective, it is necessary that they are not confined just to the level and method of charging for water but also cover a whole set of institutional and technical conditions. Current debates as to whether water pricing can influence allocation miss the fact that water pricing, like any other pricing system, is a configuration of institutional and technical conditions. Even on its cost recovery role, there are also serious issues, as the current practice of charging only the farmers is unrealistic when most of the benefits of water investments occur outside of the rural economy. Under this condition, conventional irrigation-based calculations of cost-benefit analysis of water investment underestimate its real contributions, leading to the already observed trends of underinvestment in water resources development.

Underinvestment apart, there is also the problem of scale and regional bias. The vast potential for small-scale and multiple-use water works as well as pro-poor water technologies remains unexplored as is the scope for increasing irrigation investment in regions such as Sub-Saharan Africa. As these options are inherently pro-poor and have a tremendous capacity to directly benefit the poor, there is a need for a major international thrust to step up investment on what IWMI calls the ‘soft path’ for water development. At the same time, there is also a need to sustain investment on large irrigation, especially in southern Africa. The extensive research on water-poverty linkages clearly shows that the poverty impact of irrigation investment in water, though declining over time, can be sustained with changes in land tenure, water allocation and supportive investments in farm extension and markets.

9.4.5 Water, health and environment

Even when the research shifted its focus from 'irrigation' to the larger concerns of water and its land and environmental implications, a number of key issues, which are immediately linked to irrigation but with implications far beyond its confines, still deserve closer attention. These issues are related to the health and ecological consequences of irrigated agriculture. IWMI research tries to capture these issues along with the broader issues of environmental water allocation and the ecological and livelihood functions of water-based ecosystems. Analytically, the health issues are approached in terms of the irrigation-malaria nexus, wastewater use in urban fringe agriculture, and through the role of irrigation systems in chemical contaminations and environmental flows. The ecological issues related to irrigation are considered in a functional sense and from an allocation perspective, with the former capturing the effects of irrigation on pest and vector dynamics as well as on biodiversity loss and the latter capturing the role of irrigation in supporting environmental allocation. But, the issue of environmental flow also assumes exclusive attention from the larger perspective of water-environment linkages. This fact and the fundamental linkages among poverty, health and environment make the subject crosscutting with other economic and modeling studies.

The health effects of the irrigation-malaria nexus and toxic chemicals are very serious in most of the tropical and humid irrigation systems in Asia and Africa. Irrigated systems with stagnant and poorly drained water create conditions for vector-borne infections (e.g., malaria, Japanese encephalitis and schistosomiasis), which often compound the common illnesses caused by toxic farm chemicals. What is more serious is the fact that the rural poor, especially the vulnerable groups of children and the aged, are the major victims of these health effects. Besides loss of life, health risk also leads to heavy economic costs. The magnitude of employment and income losses as well as out-of-pocket medical expenses due to malaria and other water-related health effects are high enough to make a large dent on the direct economic benefits that these groups receive from the irrigated economic system. IWMI research on malaria effects has established that the preventive costs are, in fact, lower than the curative costs and many preventive methods are related to simple techniques of cultivation, water release, irrigation application, drainage control and water sanitation in and around the irrigation systems. This clearly suggests that better irrigation and land management is essential not only for raising the levels of use efficiency and productivity of water in irrigated agriculture but also for improving the health of the poor and vulnerable rural groups. Implicit is also the point that besides the productivity and health benefits, better water and land management also has a central role in releasing more water for environmental flows.

With increasing diversion of water for the urban sector, the volume of waste and polluted water is also increasing. Although this is usually viewed as an environmental problem, it also has benefits, as wastewater is often used for irrigating urban fringe agriculture, which supports a growing number of urban and peri-urban poor, particularly in the semiarid parts of the world. IWMI estimates suggest that most of the 800 million people involved in urban fringe agriculture worldwide are relying on wastewater for irrigation and the area under this irrigation is close to a million ha around 92 cities. Under conditions where the poor lack access to regular irrigation, there is reason to treat wastewater as an asset, even though there are serious health risks. Taking a balanced position, IWMI research assesses the economic and equity benefits from this practice in relation to its health and environmental costs with a view to suggest options for increasing the former while minimizing the latter. Field research conducted in Mexico, Pakistan, India, Ghana and Vietnam show strong evidences for the higher productivity of wastewater irrigated farms and the substantial income and livelihood benefits shared by wastewater users. Although wastewater is contaminated with high chemical and bacterial levels, there were not many cases for serious health risks, as users rely on coping strategies ranging from crop choices to wastewater dilution through groundwater. These and other strategies will be the key to promote wastewater irrigation with little or no risk.

An irrigation system, especially the way it is maintained and operated, has a pivotal role both in influencing the biodiversity potential within irrigated agriculture as well as in determining the health and productivity of the downstream ecosystems. The farming practices and chemical applications in irrigated agriculture reduce biodiversity, but aggravate the problems of pests and weeds. In addition to better water management strategies, the implementation of the principles of eco-agriculture is essential to manage irrigated agriculture in a way that minimizes the conflict between agricultural production and the sustainability of freshwater-dependent ecosystems and their biodiversity. The resolution of this issue addresses only one dimension of the food-environment conflict in water management. The other and major dimension of the conflict relates to the question of water allocation between agriculture and environment. As more and more water is diverted for meeting agricultural and urban needs, the quantity and quality of flows available in many rivers are inadequate to maintain the health of the instream and downstream ecosystems. As environmental flows are also important for supporting fish production and wetland-based livelihoods, IWMI has initiated a major activity to calculate the environmental flows by basin and subbasins of major river systems in the world.

IWMI research on water productivity explores how such environmental needs can be met without affecting the water requirements for food production. Available results do suggest that environmental allocation also has a strategic role in putting pressure for agriculture to raise its productivity.

9.4.6 Water management for agriculture

Water for agriculture is inseparably linked with the question of water for the environment. As they are two parts of the same issue, there is a clear justification for them to be treated as part of the water-food-environment interface. The starting point is to determine how much water agriculture will require for meeting the growing demand for food and rural livelihoods. The answer depends clearly on water productivity, not just in irrigated agriculture but equally, if not more so, in the rain-fed system. The changing food composition also matters. With an increasing income and living standard, direct cereal consumption is declining while that of meat, dairy and fish products is increasing rapidly.

The growing significance of livestock and aquaculture, especially for the rural poor in marginal areas, also warrants water for supporting grazing areas, inland and coastal water-bodies and wetlands. Thus, improvement in water productivity and change in food composition can considerably reduce the water need for food production. At the same time, the declining share of water for agriculture need not necessarily justify the tendency either to scale down global investment in irrigation and water development or to place agriculture on the lower scale of international development priorities. On the contrary, the issue demands much more careful and delicate attention now than ever before. The CA, a larger-scale, multi-organizational research initiative led by IWMI, aims precisely to do this delicate task by looking at the issue in all its dimensions.

A number of estimates are available for the volume of water needed for agriculture by 2025. However, they vary considerably, essentially in terms of their differential assumptions on the relative output growth in irrigated and rain-fed agriculture. IWMI's own estimate suggests that irrigation would require an additional 20% water withdrawal. However, this is being now revised, as it is based on current productivity levels in irrigated agriculture and a gross underestimation of the production potential of rain-fed agriculture. In any case, considering the expansion of urban water needs and the necessity of providing more water for the environment, it is highly untenable to support additional water for irrigation, except in some regions such as Africa and Southeast Asia, where there is still scope for additional water development for irrigation. Therefore, the water, land and technology-based options for improving water productivity are critical for achieving the food, livelihood, health and environmental targets. If it is possible to raise water productivity gradually by 40% over the next two decades, then, agriculture can meet its food and livelihood demands with no additional water withdrawal by 2025. Even more promising is the vast scope for linking water productivity with water and soil conservation, particularly in rain-fed and marginal areas, which are expected to play a major role in achieving the overall productivity target. Case studies of irrigation technologies, crop choices, farming practices and water-use decisions in Central Asia, southern Africa, and South Asia suggest that this is indeed achievable with reasonable investment levels and scaling-up initiatives.

While there are trade-offs between food and environmental needs of water, attention is shifting now towards the complementarities. The food and livelihood significance of well-maintained inland water bodies, wetlands and forests, and healthy catchments and landscapes are cases in point. Even the food-environment conflict also has positive impacts, as environmental allocation can compel agriculture to raise its water productivity. These complementarities and positive spinoffs are certainly important. But, at the same time, it is essential to ensure that these effects directly address the central concerns of income generation, poverty reduction and food security. The macro issue of water allocation for food and environment has, therefore, to be addressed in conjunction with the micro issues of water access and distribution, particularly within agriculture.

IWMI research on the economic, equity and ecological impacts of irrigation development covering over a 50-year period does suggest that the poverty and income effects of irrigation investment are quite substantial and remain as important as the direct effects on food production. Although negative effects on the environment in terms of land and water salinity, catchment degradation and water pollution are growing, irrigation investments still do have positive net benefits on balance. What is needed now is a major overhaul of the water investment portfolio to target more directly on two simultaneous fronts, one on the options for improving overall water productivity and the other on pro-poor water technologies and options such as manual and miniature pumps and small-scale and multiple-use irrigation systems. The approach of balancing larger productivity goals with the poverty and equity concerns calls for much more than targeted water investments. The success in this respect depends on concurrent reforms in institutions and policies, not only those related to water but also those related to land, agriculture and environment.

9.5 KNOWDELGE BROKERING AND CAPACITY BUILDING

With a broader mandate, an integrated approach and a new perspective, IWMI research since 1996 has covered a variety of major issues falling in the interface of water, land and environment. The research results generated during this period have been significant for their alternate perspectives and new insights. The practical value of the methodological tools and scientific information developed has indeed enabled the Institute to increasingly position itself as a global knowledge center on water, food and environment. This new role has been one of the major goals set in IWMI's new Strategic Plan of 2004-08. At the same time, the capacity-building role, which is actually a key part of the organizational mandate, has also grown from sporadic training programs to more regular arrangements such as the PhD and postdoctoral fellowship programs, small grants program, and research networks with national researchers and local partners. As these formal arrangements are built within research projects, there is considerable synergy between research and capacity building.

From a larger perspective, the capacity-building activity also contributes to the knowledge role, as it is also a vehicle for the spread and application of concepts and methods in organizations and contexts beyond IWMI. The knowledge and capacity-building roles have, therefore, to be considered as an integral part of the research role itself.

9.5.1 Knowledge brokering

Knowledge generation and dissemination are conventional roles for research organizations. IWMI does generate knowledge through applied and policy-oriented research on strategically selected priority themes and geographic locations. But, besides disseminating this knowledge through the usual channels of peer-reviewed publications and policy briefs, IWMI also utilizes its unique international position to take a more proactive role in making the existing knowledge - both from its own research and that from others - available in a form that is most likely to reach the potential users such as the national and international researchers, policymakers and donors. In fact, there is now a growing demand for international organizations such as IWMI to serve as brokers and facilitators of international knowledge flow.

To play this challenging role, IWMI is also increasingly transforming itself as a learning Institute to take up the responsibility for making the knowledge to be easily accessible and available on a wider scale. This is certainly a major change for an Institute that has long been used to focus almost exclusively on routine knowledge generation and its dissemination through the normal means of research communication. This change is spearheaded both by IWMI's Information and Knowledge Group as well as by several change projects pursued within the Institute since 2000 for creating the culture of information-sharing and dissemination.

With the expanding influence of electronic media and their powerful impacts on research communication, there are new opportunities for improving knowledge presentation and dissemination. IWMI is attempting to capitalize on these new opportunities through web-based information and knowledge sharing. The key for an effective knowledge-brokering initiative is to target different groups with an appropriately packaged set of information and knowledge tools. While research outputs such as research reports, articles, books and, more importantly, methods and databases target researchers, others such as policy briefs, which present the key findings of scientific research in summary form and non-technical format, target the public, policymakers and donors. The spatial databases being developed at present, for instance, bring together information on water, food and environmental aspects from diverse sources and forms within a common platform, keeping researchers as end users.

'E-publishing' is another major activity that aims to facilitate knowledge transfer by making most of the print media-based research available in an easily sharable electronic form. Electronic versions of many IWMI publications, including its research reports, articles, proceedings, books and policy briefs can now be downloaded directly from the Institute's website. IWMI has also linked its library resources with other CGIAR centers as well as with many other national and international organizations. With these initiatives to open up the information and research resources, IWMI is trying to ensure that the same level of library and information services available to an IWMI researcher is also available to other researchers working on water, land and environmental issues around the world.

IWMI has concurrently enhanced its role in adding to scientific knowledge through the publication of peer-reviewed research reports, journal articles and books. The contributions on this front have been quite remarkable with a nearly fourfold increase in the number of publications per researcher in ISI ranked journals over the period 1997-04. Besides its regular publications, IWMI has also led the preparation of a number of special issues of journals with the express purpose of bringing together existing knowledge on some of the key themes in one set of pages usable by the international research community. Some of these special issues and special sections within regular issues are listed below.

(a) 'Research from the International Water Management Institute', *International Journal of Water Resources Development* 15 (1/2), 1999.

(b) 'Water, Poverty, and Gender', *Water Policy* 5(5/6), 2003.
(c) 'Growing More Rice with Less Water', *Paddy and Water Environment* 2 (4), 2004.
(d) 'River-Basin Management: Economics, Management and Policy', (Special Section), *Water Resources Research* 40(8), 2004.
(e) 'Malaria and Agriculture', *Acta Tropica* 89(2), 2004.
(f) 'Water Institutional Reforms: Theory and Practice', *Water Policy* 7(1), 2005.

Currently, two other special issues-one on 'Water Policy Issues in Chinese Agriculture' and the other on 'Water Productivity'-are under preparation with an arrangement with *Water Policy*, the official journal of the World Water Council. The proactive role of IWMI in knowledge dissemination is supported further with the formal publishing arrangements with leading international publishers such as CABI Publishing, Resources for the Future Press, and International Water Association Publishing. The idea behind these agreements is to supplement commercial distribution of major IWMI research with a cost-free distribution to reach research organizations and academic institutions, particularly in developing countries.

Knowledge sharing and dissemination through the organization of international workshops and conferences and presentations by IWMI researchers are also quite strong. Besides the workshops organized regularly as part of major research projects, IWMI has also organized many special international conferences on key issues such as irrigation management transfer, river basin institutions, water-poverty impacts, and water productivity. IWMI continues to be represented well in the meetings of international water-related associations as participants, resource persons and keynote speakers. For instance, in the 25th biennial conference of the International Association of Agricultural Economists, IWMI cosponsored a workshop on water institutional reforms.

Some of the special sessions and events organized by IWMI in major international meetings, such as the World Water Forums as well as the annual meetings of Stockholm Water Symposiums and Commission on Sustainable Development have been quite influential in articulating ideas such as water productivity and environmental allocations. In particular, the WaterDome event organized by IWMI during the 2002 World Summit on Sustainable Development in Johannesburg, South Africa played a major role in highlighting the centrality of water, food and environment in the global agenda for sustainable development. Thanks largely to this event and subsequent efforts, the key ideas and paradigms such as the 'more crop per drop' and 'water-food-environment nexus' are now part of the larger international development dialogue.

9.5.2 Capacity building

Since capacity building links knowledge generation with knowledge application, IWMI views capacity building as part of the knowledge-brokering process itself. The capacity-building activities of the Institute in its early years focused primarily on the training of national partners and research consultancy arrangements with local researchers. There is now a formal program for capacity-building activities, which is fully integrated into IWMI's ongoing research. Capacity building at IWMI takes a variety of forms, ranging from the intellectual and financial support for interns, masters and PhD students and postdoctoral fellows to the research alliances with national researchers and training workshops. The growing significance of capacity building can be seen from the fact that more than 130 interns, masters and PhD students and post-doctoral fellows have benefited since 2003 under IWMI's Capacity Building Program. Besides research training, this Program has also increased the number of joint research publications with developing country partners. In 2004, for instance, IWMI produced more than 100 such joint publications.

In terms of training workshops, over 100 NARES partners have received training through the Capacity Building Program since 2003. The IWMI-Tata Program on Water Policy, forming part of IWMI's India program, has organized more than 20 consultations and workshops for researchers and policymakers. In addition to these formal workshops for training and policy consultation, capacity building also occurs as an integral part of the implementation of many research projects. For example, the use of GIS-based risk maps has been introduced to malaria-control personnel in Sri Lanka and entomological expertise in taxonomy, and field sampling is presently being provided in Pakistan. IWMI has also conducted extensive capacity building of municipal authorities and local research institutes through cooperation in the testing of a pilot waste management plant (Ghana), training of masters students (Ghana), and the strengthening of the testing procedures in laboratories (Pakistan and Vietnam).

The network approach adopted under the ASIALAND and MSEC projects has formally integrated research implementation, capacity building and technology transfer in the context of soil and water conservation activities in Southeast Asia. The research results and project sites are also used to link technical training with field visits. The Africa Training Hub organized by IWMI in South Africa in 2004 for the task managers and top-level management staffs of the World Bank and national partners is an important case for linking training with knowledge-sharing and problem-solving. Another recent initiative, which falls in the interface among network creation, knowledge brokerage and capacity building, relates to the role that IWMI played in facilitating a strategic partnership between the Indian Council of Agricultural Research and the

Association for Strengthening Agricultural Research in Eastern and Central Africa. The Memorandum of Understanding was signed between the two organizations in early 2006 with an aim to support agriculture and natural resource management research and capacity development in East and Central Africa.

9.6 RESEARCH REACH AND IMPACT ASSESSMENT

Some of the ideas, concepts and tools developed by IWMI have been very influential in changing the way water issues, their impacts and solution procedures are approached and evaluated today both at the global and local levels. The idea of 'more crop per drop' and the concept of water productivity have revolutionized the way the question of use efficiency and benefits of water are reckoned. The global water scarcity maps, for instance, developed by IWMI are instrumental in countering the tendency to globalize the water crisis by locating specific regions and countries at risk. Similarly, rain-fed agriculture is viewed today as the frontier of opportunity in view of its pivotal role in meeting future food and livelihood needs, mediating water savings for environment, and balancing productivity with equity and sustainability.

IWMI has also had remarkable success in convincing policymakers on the socioeconomic importance of environmental flows as well as their strategic significance in facilitating major water productivity improvements in agriculture. The same is also the case with the argument for increasing investment on 'soft options' covering small-scale pro-poor water projects and technologies. Similarly, the socio-ecological approach to groundwater management and institutional and political economy approach to water governance have made substantial contributions to both water resources research as well as to institutional economics literature. As a learning Institute, IWMI is interested in constantly evaluating the reach and impact of its research on literature and policy. Such an evaluation is important both for the internal purposes of accountability, quality control and reorientation of research agenda as well as for the external requirements of donor agencies and the CGIAR Science Council.

9.6.1 Research reach

While the measurement of knowledge reach and impact is not an exact science, there are some commonly used approaches and proxy measures. These approaches and measures try to capture different facets evident in the long chain of the impact process involving knowledge generation, its uptake and application, and its final impacts. Understandably, the evaluation becomes more and more difficult with successive stages due to long gestation, uncertainty, and qualitative and ex-ante issues. The evaluation of initial stages is relatively easier thanks to the availability of bibliometric tools such as the Thomson Web of Science and Google Scholar as well as webmetric analyses. While these tools cannot provide complete coverage, they can serve as an indication of the demand for, and use of, major IWMI research publications.

Some preliminary analyses based on these tools suggest that some of the IWMI publications are cited more than the average publications of other comparable organizations involved in social science/policy research. For instance, a bibliometric assessment using Thomson Web of Science suggests that the IWMI Research Report *Accounting for Water Use and Productivity* was cited 16 times during 1999-02, as against an average of 4 citations per article from comparable research organizations reported in the bibliometric study of Pardey and Christian (2002). Another result based on Google Scholar indicates that of the 251 IWMI publications on irrigation management transfer, 126 were registered in the Google Scholar website and had been cited over 500 times; two-thirds of these citations are by non-IWMI researchers. An additional webmetric analysis indicated that 23 IWMI Research Reports and Working Papers on irrigation management transfer were downloaded more than 29,000 times during 2000-05. While there are a number of caveats associated with webmetric analysis, downloads tend to have a positive correlation with subsequent citations (Pinkowitz 2002; Brody and Harnad 2005) and hence, can serve as an early indicator of research reach and use.

9.6.2 Research application

While bibliometric analysis does suggest that the knowledge that IWMI generates is increasingly utilized in the academic literature, the ultimate goal for IWMI is to ensure that its research results are actually applied to have a positive influence on food production, livelihood generation and environmental sustainability. Achieving this goal is perhaps the most difficult and enduring challenge for IWMI. It requires the formation of strong links with the scientific and policy communities as well as donor and development agencies. It also needs constant monitoring and evaluation of the pathways linking research, its uptake, application and impact. Since the effectiveness of these pathways depends, among other things, on the appropriateness, applicability and potential impact of research results, IWMI places considerable emphasis on stakeholder needs and feedback from field application as mechanisms for aligning its research agenda and thematic priorities with the changing conditions. Since the assessment of the application and impact of research is critical for the mission of IWMI, impact assessment has now become an integral part of the Institute.

Here, a few illustrative cases for the adoption and application of knowledge and practices generated by IWMI are provided. They cover both the adoption of broad ideas as well as the application of specific results.

As indicated elsewhere, the 'more crop per drop' is underlying what the United Nations calls the 'Blue Revolution'. Similarly, the concept of water productivity and the method of water accounting have now become the standard tools for productivity evaluation and resource assessment. IWMI's research on river basin institutions and irrigation-poverty linkages has not only influenced the development agenda of the Asian Development Bank (ADB) but also led to the creation of some important initiatives. For instance, following the completion of IWMI's ADB-funded study on river basin institutions in Asia, the ADB launched a network of Asian river basin organizations, of which IWMI is also a member.

Similarly, the guidelines for using irrigation for poverty alleviation developed from the results of IWMI research on the subject were not only endorsed by the ADB but are also being used to guide its irrigation investment decisions. On this count, IWMI has become a founding partner of the ADB's Water and Poverty Initiative. The success of IWMI research on eco-agriculture conducted in southern Sri Lanka has convinced the irrigation authorities and engineers to plan for the conservation of identified biodiversity hot spots within an irrigation development area and also to set aside a special area for development as an arboretum. Also in Sri Lanka, the use of IWMI's parsimonious hydrological model, which predicted flooding of a nearby city following a planned extension to an irrigation scheme, prompted the irrigation agency to make the necessary adaptations in the design of the extension area.

Several examples are available for the employment of IWMI tools, technologies and techniques at local and national levels. For instance, the guidelines for sloping land agriculture developed as part of IWMI's catchment management research have been adopted in national manuals and, even in legislation in Southeast Asia. On a local scale, with the help of NGOs such as CARE Philippines, some of the successful technologies to conserve soil in erosion-prone areas identified in IWMI's pilot sites are now spreading to other villages in the country.

The Bright Spots research, although relatively new, is also generating considerable interest in Thailand. In 2002, IWMI, along with the researchers from the Khon Kaen University and farmer networks, began to examine low-cost, ecologically sustainable, and locally accepted technologies to reverse soil degradation in Northeast Thailand. The research focused on the potential role of bentonite clays in rejuvenating soils as an alternative to the current but unsustainable use of termite mounds and dredged materials from reservoirs. Since the project's inception, at least 500 farming families in 200 villages in

Northeast Thailand have been directly benefited from the adoption of this technology with substantial economic and sustainability gains. Notably, the Land Development Department of the Government of Thailand has adopted this technology and presently, the Department is trying to enhance the coverage of the technology.

One of the more influential outputs from IWMI's wastewater program has been the 'Hyderabad Declaration', which was a major outcome of an IWMI-IDRC Conference attended by 47 international experts in wastewater research. This widely translated and disseminated Declaration calls for accepting the realities of wastewater use and its food, livelihoods and health implications in the context of poor countries. The underlying principles of this Declaration and related IWMI research on the subject are being used in several key public health guidelines, such as the *Guidelines for Water Reuse* (USEPA/USAID 2004) and the *Guidelines for the safe use of wastewater, excreta and grey water: Wastewater use in agriculture* (WHO, 2006). Another instance of policy-level impact relates to IWMI's research on the analysis of the health hazards of pesticides in irrigated agriculture. This research was one of the catalysts for the launching of a Presidential Task Force on pesticide abuse in the country. Further, IWMI research on this issue has also led to the call for an international declaration on the 'minimum pesticides lists' that would limit the range and availability of toxic chemicals.

In an important sense, the increasing demand for IWMI's involvement in many regional and global development initiatives and policy conventions also indicates the growing influence of IWMI's research knowledge. For instance, both state and national governments of India, Pakistan and Cambodia have requested IWMI's involvement in action research, policy development and training activities related to participatory irrigation management. Recognizing its contributions to the existing body of knowledge on small-scale water technologies and their application in biophysical and socioeconomic contexts, IWMI has been invited as a key partner in the Smallholder Irrigation Market Initiative (SIMI 2003) and the PRODWAT Thematic Group (IRC 2004). IWMI has also been asked to provide research and technical support for NEPAD, the World Bank and the African Development Bank, particularly in scaling up small-scale irrigation technologies and land management strategies. Similarly, thanks to the influential work on the ecological aspects of agricultural systems, IWMI is now confirmed as the fifth International Organization Partner to the Ramsar Convention on Wetlands.

9.6.3 Impact assessment

Traditionally, research organizations have concentrated on knowledge generation and its dissemination with the assumption that the expected economic benefits from the application of such knowledge will lead to its uptake and application by policymakers and development planners. This assumption is obviously unrealistic, as the application of potentially useful knowledge is constrained often by both technical and political economy constraints. In its knowledge role, IWMI strongly believes in the vast scope for a more proactive role in promoting knowledge application, particularly by liaising better with researches, policymakers and development agencies. One powerful tool to perform this vital role is impact assessment, as it can demonstrate and convince the likely benefits to policymakers, donors and local users. Besides this outreach role, impact assessment also has many functional and strategic roles. It helps to monitor the achievement of IWMI's missions in the context of specific projects, impart an impact culture within the Institute, ensure accountability to donors and funding agencies, and align the research agenda well with policy changes and stakeholder feedback. It is in view of such critical roles that impact assessment assumes a central role in the operation and functioning of IWMI as a knowledge Institute. While it is desirable to assess the impacts of all projects, for practical reasons, impact assessment is confined only to major projects, with an aim to cover, at least, two projects per year.

To better institutionalize the process of impact assessment, IWMI launched the impact assessment initiative in 2002. While still evolving, this initiative has expanded considerably over the past three years with the development of a generic framework, formal procedures and guidelines as well as the completion of pilot ex-post impact studies of projects and programs. The generic framework providing the conceptual basis for impact assessment at IWMI (see Giordano 2003) is depicted in Figure 9.1. The hallmark of this framework is its emphasis on identifying impact pathways for projects and programs while tracking and measuring progress through intermediary outcomes. These impact pathways explicitly cover all the major outcomes ranging from awareness creation to actual field-level impacts and implicitly link research partners and intermediaries involved in the entire process. There are inherently long and variable time lags among research, its uptake and final impact. Clearly, it is not easy to delineate and measure the impacts created by a particular project, especially given the role of exogenous factors and the multi-scalar nature of the impacts. However, tracking project outcomes, as much as possible, using a variety of qualitative and quantitative means, could still be very valuable to anticipate and assess the general direction of project impacts. This ex-ante dimension of impact assessment is as important as the ex-post analysis of impacts, as it allows mid-course adjustments and contributes, thereby, to the ex-post impact itself.

IWMI, like all international policy research organizations, plans its projects with the aim of generating lasting and wider impacts on water, land and environmental management in terms of food, livelihood and sustainability gains. But, the nature and extent of these impacts are determined by the impact on many intermediary pathways - both direct and indirect - operating between research and its final impact. These pathways, as per the IWMI impact typology, are:

a) Raised awareness,
b) Creation of new knowledge,
c) Application of improved tools, technologies and techniques,
d) Enhanced capacity,
e) Strengthened partnerships,
f) Adoption of improved policies/institutions, and
g) Actual impact on food, livelihood and environment.

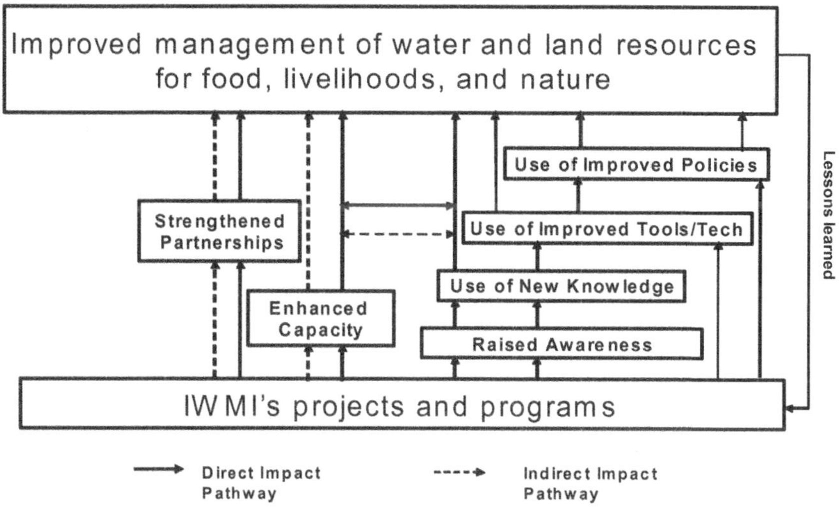

Figure 9.1 IWMI impact typology schematic.

Each of these pathways requires IWMI to interact with different players such as the publi, research community, partners, policymakers and users. But, as we go down the list, the ability of IWMI to have a direct influence on the players becomes more difficult and challenging. Even though the final impacts will be delayed and difficult to measure, the impacts on the intermediary pathways can be broadly assessed using various indicators and measurement tools. While the current focus of the typology is on individual projects, progress toward the achievement of these intermediary impacts will help IWMI and its partners better gauge their contributions toward the ultimate goal of making a major difference in the life of the poor and in the health of water-based ecosystems.

Although the impact assessment typology is still evolving, it has been used, on a pilot basis, in a series of ex-post impact assessment studies being conducted at IWMI since 2002. The objectives of these studies are to gauge the outcomes and impacts of specific IWMI programs and projects and also to test a variety of methods and tools for measuring different impact pathways. So far, three pilot studies have been conducted to evaluate the impacts of major research programs and projects. These include the impact assessment of IWMI research programs on irrigation management transfer (Giordano *et al.* forthcoming), IWMI's water accounting methodology, and a specific project on water management for malaria control in Sri Lanka. IWMI has also conducted impact assessments of two soil and water conservation projects, i.e., MSEC (Maglinao *et al.* 2003) and ASIALAND (Maglinao *et al.* 2005).

Based on the experience gained so far, some important changes have already occurred or are being contemplated on the method and modus operandi of impact assessment. Important among them are included the exploration for refining impact assessment with the adoption of new approaches and unconventional methods such as the 'outcome mapping approach' and improved procedures for accounting for stakeholders' perception in general and users' expectations in particular. The 'outcome mapping approach' is very important to capture the changing behaviors and perceptions of its 'boundary partners', i.e., defined as "those individuals, groups and organizations with whom a project team interacts directly to effect change and with whom the project can anticipate some opportunities for influence" (IDRC 2005, 2).

As impact assessment is now a common goal for the CGIAR centers and other organizations, IWMI is both formally networking with them through the CGIAR Standing Panel on Impact Assessment as well as informally sharing experiences and ideas on the relatively challenging area of assessing the impact of natural resources management. A key lesson learned is the need to involve appropriate development partners (NARES, academic institutions, local NGOs and international development agencies) early in the research process, not only to assist in designing research projects but also to identify the intended outcomes and impacts and pathways towards their achievement. As this is now followed in several IWMI projects and programs, including the Bright Spots Project and the IWMI-Tata Water Policy Program, it is likely to facilitate a better analysis of ex-post impact.

9.7 LOOKING FORWARD

As the extent and impacts of the research and knowledge contributions of IWMI over the past decade are reflected against its strategic plans and organizational mandates, certainly, there are remarkable achievements. It has adjusted relatively well with its broader mandates and new thematic structures and responded most effectively to the changing nature of the global water challenge. With a concerted effort and strong commitment, IWMI has played a leading role both in presenting the correct picture on the water question and in articulating it more firmly into the international research and development agenda. With its information and knowledge initiatives, it is emerging, gradually but clearly, as a leading international knowledge center on water, food and environment. With its strategic investments on capacity building, knowledge brokering and impact assessment, IWMI is constantly striving to enhance the value and utility of its research to the scientific community, the policymakers and resource users. Much of this is possible due to the strong support that IWMI is getting from its donors and partners. At the same time, the internal changes effected within the Institute are a driving force behind the Institute's aim to continually improve its research performance and influence.

Besides the process-related changes such as project monitoring, impact assessment and quality control, there are also major structural changes in the staffing pattern, research organization and the management structure. The performance-based system of staff classification and a decentralized system of decision making have, in particular, contributed to an environment of incentives and performance. Within the CGIAR system, IWMI was, in fact, the first to surpass its gender and regional diversity goals. While these changes were important in streamlining the performance and resilience of the Institute, the most important change, however, has occurred on the research front, particularly to make the research agenda and research operation more responsive to stakeholders and end users. This had a major impact on the research agenda. As the broader research mandate that IWMI has assumed since 1996 led to a broader research canvas, there is a need for having a clear focus on what themes and issues can be addressed most effectively with available resources and within a given time. The priority themes, focal regions, and benchmark basins were used in the recent refinement of IWMI's research focus. More importantly, IWMI's focus on water productivity has enabled IWMI research to bring out more directly the role that water and land management issues can play in poverty alleviation and environmental sustainability.

While the regional and basin focus continues to remain the same, the thematic priorities have been adjusted over time to respond to changing conditions and understandings. The five themes that have been pursued since

2000 and described in detail in this volume were reorganized in 2005 to address the way water productivity and water access could improve livelihood needs, environmental allocations and resource conservation. Now, there are four core research themes: Basin Water Management; Land, Water, and Livelihoods; Agriculture, Water, and Cities; and Water Management and Environment. These themes are not entirely new but result from a careful reorganization and refinement of the earlier themes, based on the recommendations of an external review of IWMI and through a process of consultations with IWMI researchers and stakeholders.

Of the five previous themes, three are retained with considerable reorganization and refinement while the other two (Sustainable Groundwater Management and Water Resources Institutions and Policies) are merged across the four new themes due to their crosscutting nature. Thus, the new themes have retained, more or less, the core issues of all the previous themes, but direct research more sharply on the interlinked process of productivity, allocation and access and their implications for food, livelihood and sustainability. This allows IWMI to maintain a healthy balance between change and continuity. While IWMI has made significant progress over the past decade, like most international research organizations, it still has a long way to go in realizing its goals. But, with the substantial changes in the organization, a new thematic focus and impact culture, and continuing support of its partners and donors, IWMI is now well placed to look forward to make further progress in achieving its mandate of delivering high-quality knowledge on water, food and environment to the global water community.

References

Abayawardana, S., Raschid-Sally, L., Ensink, J. H. J., van der Hoek, W. & Scott, C.A. 2001, Reuse of urban wastewater in agriculture: Current practices and issues. Paper presented at the GWP-SASTAC Workshop on "Reuse of Treated Wastewater and Sludge for Agriculture in South Asia. 7-8 December 2001". Pune, India.

Abdel-dayem, S., Hoevenaars, J., Mollinga, P.P., Scheumann, W., Slootweg, R. & Steenbergen, F. van. 2004, Reclaiming Drainage: Towards an Integrated Approach. Agricultural and Rural Development Report 1. World Bank, Washington D.C.

Abernethy, C.L., Sally, H., Lonsway, K. & Maman, C. 2000, Farmer-Based Financing of Operations in the Niger Valley Irrigation Schemes. IWMI Research Report 37. Colombo, Sri Lanka, International Water Management Institute.

ADAPT. 2004, *Climate Change in Contrasting River Basins: Adaptation Strategies for Water, Food and Environment*, edited by J. Aerts and P. Droogers. CABI HB 0851998356

Aerts J. (ed.). 2004, Adaptation Strategies to Climate Change in Contrasting River Basins: Water, Food, Environment. CABI Monograph.

Afrane, Y.A., Klinkenberg, E., Drechsel, J.P., Owusu-Daaku, K., Garms, R. & Kruppa, T. 2003, Does irrigated urban agriculture influence the transmission of malaria in the city of Kumasi, Ghana? Acta Tropica 89, 125-134.

Agarwal, A., Sunita, N. & Khurana, I. 2001, Making Water Everybody's Business: Practice and Policy of Water Harvesting. New Delhi, Centre for Science and Environment.

Albinson, B. & Perry, C.J. 2002, Fundamentals of Smallholder Irrigation: The Structured System Concept. IWMI Research Report 58. Colombo, Sri Lanka, International Water Management Institute.

Alcamo, J., Henrichs T. & Rosch, T. 2000, World water in 2025: Global modeling and scenario analysis. In: *World Water Scenarios Analyses*, edited by Rijsberman. London, Earthscan Publications.

Allan, T. 1998, Moving water to satisfy uneven global needs: "Trading" water as an alternative to engineering it. ICID Journal 47(2), 1-8.

Amerasinghe. 2004, Irrigation and mosquito-borne diseases. Journal of Parasitology (Special edition: Selected papers of the 10th International Congress of Parasitology). (in press).

Amerasinghe, F.P., Tharme, R. & Bambaradeniya, C.N.B. 2003, Eco-agriculture: A new paradigm for food and environmental security? [Abstract]. In: *Proceedings of a National Workshop on Water for Agriculture and Rural Development in Sri Lanka. Ministry of Irrigation and Water Management, Food and Agriculture Organization, and The Japanese Institute of Irrigation and Drainage.*

Amerasinghe, F.P., Konradsen, F., Fonseka, K.T. & Amerasinghe, P.H. 1997, Anopheline (Diptera: culicidae) breeding in a traditional tank-based village ecosystem in Northcentral Sri Lanka. Journal of Medical Entomology 34, 290-297

Amerasinghe, F.P., Konradsen, F., van der Hoek, W., Amerasinghe, P. H., Gunawardena, J.P.W. & Fonseka, K.T. 2001, Small Irrigation Tanks as a Source of Malaria Mosquito Vectors: A Study in Northcentral Sri Lanka. IWMI Research Report No. 57. Colombo, Sri Lanka, International Water Management Institute.

Amerasinghe, F.P., Matsuno, Y., de Silva, R.I. Piyankarage, S. C., Bambaradeniya, C.N.B. & Mallawatantri, A. 2002, Impact of irrigation on water quality, fish and avifauna of three coastal lagoons in southern Sri Lanka. 9 pp. In: *The Asian Wetlands: Bringing Partnerships into Good Wetland Practices. Proceedings of the Asian Wetland Symposium, Penang, Malaysia, 27-30 August 2001*, edited by Ahyaudin Ali *et al.* Penerbit Universiti Sains Malaysia.

Amerasinghe, P.H., Amerasinghe, F.P., Konradsen, F., Fonseka, K.T. & Wirtz, R.A. 1999, Malaria vectors in a traditional dry zone village in Sri Lanka. American Journal of Tropical Medicine and Hygiene 60, 421-429.

Annan, K.A. 2000, Chapter V: Sustaining our future. In: *We the Peoples: The Role of the United Nations in the 21st Century.* Report of the UN Secretary-General to the Millennium Summit. A/54/2000. p. 8.

Asghar, M. N., Prathapar, S. A. & Shafique, M. S. 2002a, Extracting relatively-fresh groundwater from aquifers underlain by salty groundwater. Agricultural Water Management 52 (2), 119-137.

Asghar, M. N., Saeed, M. M., Ashraf, M. & Qureshi, A. S. 2002b, Potential of skimming wells for extracting relatively-fresh groundwater from unconfined aquifers underlain by saline groundwater. *Proceedings of the 2nd International Conference on New Trends in Water and Environmental Engineering for Safety and Life: Eco-compatible Solutions for Aquatic Environments.* Capri, Italy, 24-28 June 2002.

Auerbach, R. 1999, Design for participation in ecologically sound management of S. Africa's Mlazi River catchment. PhD thesis, Netherlands, Wageningen University. 260 pp.

Badiger, S. 2003, Poverty Focused Smallholder Water Management Systems: Promoting Innovative Water Harvesting and Irrigations Systems to Support Sustainable Livelihoods in S. Asia. Final Report to DfiD. IWMI, May 2003. Bangkok. Ecological Modeling, 139 (1), 63-74.

Bahri, A. 2002, Integrated Management of Limited Water Resources in Tunisia - Research Topics Priorities, EU Meeting, Marrakech (Morocco), 16-19 May 2002, 19 p.

Bakker, M. & Matsuno, Y. 2001, A framework for valuing ecological services of irrigation water. Irrigation and Drainage Systems 15, 99-115.

Bakker, M., Barker, R., Meinzen-Dick, R. & Konradsen, F. (eds.). 1999, Multiple Uses of Water in Irrigated Areas: A Case Study from Sri Lanka. SWIM Paper No. 8. Colombo, Sri Lanka, International Water Management Institute.

Ballabh, V. & Chowdhury, K. 2003, Groundwater development and agriculture production: A comparative study of Eastern UP, Bihar & West Bengal. Anand, India: IWMI-Tata Water Policy Program. Unpublished paper.

Bambaradeniya, C. N. B. & Amerasinghe, F.P. 2002, Biodiversity assessment of the Udawalawe irrigation extension development project area: Summary report of predevelopment status of biodiversity in the project area. Report submitted to the Mahaweli Authority of Sri Lanka. 08 pp. (Unpublished).

Bambaradeniya, C. N. B. & Amerasinghe, F. P. 2003, Biodiversity Associated with the Rice Field Agro-ecosystem in Asian Countries: A Brief Review. Working Paper 63. Colombo, Sri Lanka, International Water Management Institute.

Bandaragoda, D.J. 1998a, Design and Practice of Water Allocation Rules: Lessons from Warabandi in Pakistan's Punjab. IWMI Research Report 17. Colombo, Sri Lanka, International Water Management Institute.

Bandaragoda, D.J. 1998b, Need for Institutional Impact Assessment in Planning Irrigation System Modernization. IWMI Research Report 21. Colombo, Sri Lanka, International Water Management Institute.

Bandaragoda, D.J. 2000, A Framework for Institutional Analysis for Water Resources in a River Basin Context. IWMI Working Paper 5. Colombo, Sri Lanka, International Water Management Institute.

Barbier, E. B. & Thompson, J. R. 1998, The value of water: Floodplain versus large-scale irrigation benefits in Northern Nigeria. Ambio 27(6), 334-440.

Barker, R. & Molle, F. 2002, Perspectives on Asian irrigation. Paper presented at the Conference on Asian Irrigation in Transition-Responding to the Challenges Ahead. Asian Institute of Technology, Bangkok, Thailand, 22-23 April 2002. 30p.

Barker, R. & Molle, F. 2003, Evolution of Irrigation in Asia. Comprehensive Assessment of Water Management in Agriculture. Draft Research Report. Colombo, Sri Lanka, International Water Management Institute.

Barker, R. & Molle, F. 2004, Evolution of Irrigation in South and Southeast Asia. Research Report 5, Colombo, Sri Lanka, Comprehensive Assessment Secretariat.

Barker, R., Dawe, D. & Inocencio, A. 2003, Economics of water productivity in managing water for agriculture. In: *Water Productivity in Agriculture: Limits and Opportunities for Improvement,* edited by J. W. Kijne, R. Barker and D. Molden. Wallingford, UK, CABI Publishing.

Barker, R. & Herdt, R. W. with Rose, B. 1985, The Rice Economy of Asia. Washington, D.C., Resources for the Future.

Barker, R., van Koppen, B. & Shah, T. 2000, A Global Perspective on Water Scarcity and Poverty: Achievements and Challenges for Water Resource Management. Colombo, Sri Lanka, International Water Management Institute.

Barker, R., Loeve, R. Li, Y.H. & Tuong, T.P. (eds.) 2001, Water-Saving Irrigation for Rice: *Proceedings of an International Workshop on Water Saving Irrigation in Paddy Rice Held in Wuhan, China, 23-25 March 2001.* Colombo, Sri Lanka, International Water Management Institute.

Bastiaanssen, W. G. M. 1998, Remote Sensing in Water Resources Management: The State of the Art. Colombo, Sri Lanka, International Water Management Institute.

Bastiaansen W., Molden, D., Thiruvegadachari, S., Smit, A., Mutuwatte, L. & Jayasinghe, G. 1999a, Remote Sensing and Hydrologic Models for Performance Assessment in Sirsa Irrigation Circle, India. IWMI Research Report 27. Colombo, Sri Lanka, International Water Management Institute.

Bastiaanssen, W. G. M., Thiruvengadachari, S., Sakthivadivel, R. & Molden, D. J. 1999b, Satellite remote sensing for estimating productivities of land and water. International Journal of Water Resources Development 15(1/2):181-196.

Batchelor, C., Cain, J., Farquharson, F. & Roberts, J. 1998, Improving Water Utilization from a Catchment Perspective. IWMI SWIM Paper 4. Colombo, Sri Lanka, International Water Management Institute.

Bauer, C. J. 1997, Bringing water markets down to earth: The political economy of water rights in Chile. World Development 25(5).

Bauer, C. J. 1998, Slippery property rights: Multiple water uses and the neoliberal model in Chile, 1981-95. Natural Resources Journal l38 (1), 109-155.

Bhalla, G. S. & Singh, G. 2001, Indian Agriculture: Four Decades of Development. New Delhi, Sage Publications.

Bhamoria, V. 2002, Wastewater and welfare: Pump irrigation economy of peri-urban Vadodara. Anand, India: IWMI-Tata Water Policy Program. Unpublished paper.

Bhatt, Y. & Vallee, D. 2004, The Global Dialogue on Water, Food and Environment: Lessons Learnt and Looking Ahead. Unpublished Final Report. July 2004. Colombo, Sri Lanka, Dialogue Secretariat.

Bhattarai, M. & Narayanamoorthy, A. 2003, Impact of irrigation on rural poverty in India: An aggregate panel-data analysis. Water Policy 5 (5/6), 443-458.

Bhattarai, M., Barker, R. & Narayanamoorthy, A. 2003, Who benefits from irrigation development in India? Implication of irrigation multipliers for cost recovery, and irrigation financing. In: *ICID Asian Regional Workshop, Sustainable Development of Water Resources and Management and Operation of Participatory Irrigation Organizations, November 10-12, 2003, The Grand Hotel, Taipei.* Vol.1. ICID, Taipei, Taiwan. pp.285-296.

Bhattarai, M., Pant, D., Mishra, V. S., Devkota, H., Pun, S., Kayastha, R. N. & Molden, D. 2002, Integrated Development and Management for Water Resources for Productive and Equitable Use in the Indrawati River Bain, Nepal. IWMI Working Paper 41. Colombo, Sri Lanka, International Water Management Institute.

Bin, R.D., Loeve, Y., Li, C. C., Li, Deng & Molden, D.J. 2001, Water productivity in Zhanghe irrigation system: Issues of scale. In: *Proceedings of the International Workshop on Water Saving Irrigation for Paddy Rice, 23 – 25 March 2001, Wuhan,* edited by R. Barker, R. Loeve, Y.H. Li and T.P. Tuong. Colombo, Sri Lanka, SWIM publication.

Boelee, E. & Laamrani, H. 2003, Multiple use of irrigation water in northern Morocco. In: Proceedings of the International Symposium on water, poverty, and productive uses of water at the household level, 21-23 January 2003, Muldersdrift, South Africa, edited by J.A. Butterworth, P. B. Moriarty, M. Venter Hildebrand, B. van Koppen, B. Schreiner and D. Versfeld. Natural Resources Institute, IRC International Water and Sanitation Centre, Department of Water Affairs and Forestry, International Water Management Institute, p. 35-43.

Boelee, E. & Laamrani, H. 2004, Multiple use of irrigation water in Northeastern Morocco. In: Beyond Domestic: Case Studies on Poverty and Productive Uses of Water at the Household Level, edited by P. Moriarty, J. Butterworth and B. van Koppen. Delft, Netherlands, IRC International Water and Sanitation Centre; International Water Management Institute. pp.119-135.

Boelee, E. & van der Hoek, W. 2002, Impact of Irrigation on Drinking Water Availability in Sri Lanka (Impact de L'irrigation Sur la Disponibilité de L'eau Potable au Sri Lanka). ICID-CIID 18th Congress on Irrigation and Drainage, 21-28 July, Montreal, Canada. Q. 51, R. 5.04. International Commission on Irrigation and Drainage.

Boelee, E., Amerasinghe, P. & Amerasinghe, F. 2003, Water management for malaria control in an area of the Huruluwewa watershed. Final Report, submitted to the Ministry of Health, Government of Sri Lanka. 11 pp.

Boelee, E., Laamrani, H. & van der Hoek, W. 2000, Multiple use of irrigation water in dry regions of Africa and south-Asia. International Conference "Water and Health – Ouaga 2000. Health and Nutritional Impacts of Water Development Projects in Africa", Ouagadougou, Burkina Faso, 21-24 November 2000. Session 2: 1-6.

Boisvert, R., Chang, H., Barker, R., Levine, G., Matsuno, Y. & Molden, D.J. 2003, Water productivity in agriculture: Measuring the positive and negative externalities of paddy rice production. Presented at the Third World Water Forum, Kyoto, Japan; forthcoming IWMI Research Report.

Boonsaner, A., Lorent, C. & Janeau, J. L. 2002, Off-site effect of erosion: The silting of Mae Thang reservoir. Proceedings of the 7th MSEC Annual Assembly, 2002. Colombo, Sri Lanka, International Water Management Institute.

Bradley, D. 1977, The health implications of irrigation schemes and man-made lakes in tropical environments. In: *Water, Wastes and Health in Hot Climates*, edited by R. Feachem, M. Mcgarry and D. Mara. John Wiley, London, p. 18-29.

Breman, J.G. 2001, The ears of the hippopotamus: Manifestations, determinants, and estimates *of* the malaria burden. American Journal of Tropical Medicine and Hygiene 64 1, Suppl., 1-11.

Brewer, D. J., Sakthivadivel, R. & Raju, K.V. 1997, Water Distribution Rules and Water Distribution Performance: A Case Study in the Tambraparani Irrigation System. IWMI Research Report 12. Colombo, Sri Lanka, International Water Management Institute.

Bridges, E.M., Hannam, I.D., Oldeman, L.R., Penning de Vries, F.W.T., Scherr, S.J. & Sombatpanit, S. (eds.). 2001, Response to Land Degradation. USA, Science Publishers. 510 pp.Bakker, Margaretha, Barker, Randolph, Meinzen-Dick Ruth, & Konradsen, Flemming. 1999, Multiple Uses of Water in Irrigated Areas: A Case Study from Sri Lanka. IWMI SWIM Paper 8. Colombo, Sri Lanka, International Water Management Institute.

British Geological Survey. 2004, Community Management of Groundwater Resources: An Appropriate Response to Groundwater Overdraft in India? London, British Geological Survey.

Brody, T. & Harnad, S. 2005, Earlier Web Usage Statistics as Predictors of Later Citation Impact. Technical Report, Electronics and Computer Science School, University of Southampton. (http://eprints.ecs.soton.ac.uk/10712)

Bruns, B. & Bandaragoda, D.J. (eds.). 2003, Governance for Integrated Water-Resources Management in a River Basin Context. *Proceedings of the Regional Seminar, Bangkok, May, 2002.* Colombo, Sri Lanka, International Water Management Institute. 264pp

Bruns, B., Bandaragoda, D.J. & Samad, M. (eds.). 2002, Integrated Water-Resources Management in a River Basin Context: Institutional Strategies for Improving Productivity of Agricultural Water Management. *Proceedings of the Regional Workshop, Malang, Indonesia, January 15-19, 2001.* Colombo, Sri Lanka, International Water Management Institute, 315pp

Buechler, S. & Devi, G. M. 2003, The Impact of Water Conservation and Reuse on the Household Economy. *Proceedings of 8th International Conference on Water Conservation & Reuse of Wastewater.* Mumbai, 13-14 September 2003.

Buechler, S. & Devi, G. 2003a, Wastewater as a source of multiple livelihoods? A study of a rural area near Hyderabad city, Andhra Pradesh, India. pp 939-948. In: Water and Wastewater: Developing Country Perspectives, edited by Rema Devi and Naved Ahsan. London, UK, International Water Association.

Buechler, S. & Devi, G. 2003b, Household food security and wastewater-dependent livelihood activities along the Musi river in Andhra Pradesh, India. In: WHO Guidelines for Wastewater Use for Agriculture. Vol II. (invited paper, submitted)

Buechler, S. & Scott, C. 2000, "Para nosotros, esta agua es vida". El riego en condiciones adversas: Los usuarios de aguas residuales en Irapuato, México ["For us, this water is life". Irrigation under adverse conditions: wastewater users in Irapuato, Mexico]. In: Asignación, Productividad y Manejo de Recursos Hídricos en Cuencas, edited by

C. Scott, P. Wester and B. Marañón-Pimentel. Serie Latinoamericana. Guanajuato, Mexico: International Water Management Institute.

Buechler, S. & Zapata, E. (eds.). 2000, Género y Manejo de Agua y Tierra en Comunidades Rurales de México (Gender and the Management of Water and Land in Mexico). Serie Latinoamericana. Mexico, D.F., International Water Management Institute; Montecillo, the Colegio de Postgraduados.

Buechler, S., Devi, M. G. & Raschid-Sally, L. 2002, Livelihoods and wastewater irrigated agriculture along the Musi river in Hyderabad city, Andhra Pradesh, India. Urban Agriculture Magazine 8, 14-17.

Burke, J. & Moench, M.. 2000, Groundwater and Society: Resources, Tensions and Opportunities. New York, The United Nations.

CABI Publishing (in association with IWMI). 2003, Water Productivity in Agriculture: Limits and Opportunities for Improvement, edited by J. W. Kijne, R. Barker and D. Molden. Comprehensive Assessment of Water Management in Agriculture Series. 332 p.

Cai, X. & Rosegrant, M. 2003, World water productivity. In: Water Productivity in Agriculture: Limits and Opportunities for Improvement, edited by J.W.Kijne, R. Barker and D. Molden. Wallingford, UK; Colombo, Sri Lanka: CABI; IWMI.

Cairncross, S. 1997, More water: Better health. People and Planet 6, 3.

Calder, I. R. 1998, Water-Resource and Land-Use Issues. SWIM Paper 3. Colombo, Sri Lanka, International Water Management Institute.

Campbell, B.M., Jeffery, S., Kozanayi, W., Luckert, M., Mutamba, M. & Zindi, C. 2002, Household Livelihoods in Semi-Arid Regions. Bogor, Indonesia, CIFOR. 153 pp.

Challenge Program on Water and Food Consortium. 2002, Full Proposal. Colombo Sri Lanka.

Chambers, R. 1988, Managing Canal Irrigation: Practical Analysis from South Asia. New Dlhi, India, Oxford and IBH Publishing Co. Pvt. Ltd. 279pp.

Chambers, R., Saxena, N.C. & Shah, T. 1987, To The Hands of the Poor: Water and Trees. London, UK, Intermediate Technology Publications.

Chaney, R.L., Ryan, J.A., Reeves, P.G. & Simmons, R.W. 2002, A new paradigm for environmental cadmium risk assessment. 2nd Conference on Molecular Mechanisms of Metal Toxicity and Carcinogenesis September 8th -11th, 2002 NIOSH, Morgantown, WV, USA.

Chaplot, V., Giboire, G., Marchand, P. & Valentin, C. 2004, Dynamic modeling for gully prediction under global change in northern Laos. Catena (submitted).

Chaplot, V., Boonsaner, A., Bricquet, J. P., de Rouw, A., Janeau, J. L., Marchand, P., Phommasack, T., Toan, T. D. & Valentin, C. 2002, Soil erosion under land use change from three catchments in Laos, Thailand and Vietnam. *Proceedings ISCO Meeting* IV, 313-318, Beijing, China.

Christen, E. & Van Meerveld, I. 2000, Institutional Arrangements in the Shepparton Irrigation Region, Victoria, Australia. IWMI Pakistan Report No.R-99. Pakistan National Program. Lahore, Pakistan, International Water Management Institute, v, 64p.

Cosgrove, W.J. & Rijsberman, F.R. 2000, World Water Vision: Making Water Everybody's Business. London, Earthscan Publications.

Coughlan, K.J. & Lefroy, R.D.B. 2001, Assessment of Sustainable Land Management Using the FESLM: The IBSRAM Experience. Issues in Sustainable Land Management 8. Bangkok, International Board for Soil Research and Management. 67 pp.

Coward, E.W. Jr. 1986, Management themes in community irrigation systems. In: *Irrigation and Agricultural Development in Asia: Perspectives from the Social Sciences,* edited by E.W. Coward, Jr. Ithaca, New York, USA, Cornell University Press. pp. 203-218.

Craswell, E. 1998, A decade of progress at IBSRAM and future priorities. In: Soil Quality and Agricultural Sustainability, edited by R. Lal. Chelsea, Michigan, Ann Arbor Press. pp. 337-348.

Craswell, E.T. & Niamskul, C. 2000, Watershed management for erosion control in steeplands in Asia. In: Integrated Watershed Management in the Global Ecosystem, edited by R. Lal. Boca Raton, FL., CRC Press. pp 65-72.

Craswell, E.T., Sajjapongse A., Howlett, D.J.B. & Dowling, A.J. 1997, Agroforestry in the management of sloping lands in Asia and the Pacific. Agroforestry Systems 38, 121-137.

Custodio, E. & Llamas, M. R. 2003, Intensive use of groundwater: Introductory considerations. In: Intensive Use of Groundwater: Challenges and Opportunities, edited by R. Llamas and E. Custodio. Rotterdam, Netherlands, A.A. Balkema.

Danso, G., Drechsel, P., Wiafe-Antwi, T. & Gyiele, L. 2002, Income of farming systems around Kumasi. Urban Agriculture Magazine 7, 5-6.

de Fraiture, C., Molden, D., Amarasinghe, U. A. & Makin, I. 2001, PODIUM: Projecting water supply and demand for food production in 2025. Phys. Chem. Earth 26, 11-12, 869-876. Elsevier Science.

de Rouw, A. 2001, Weed infestation and soil erosion resulting from the breakdown of the slash and burn cultivation system. In: Soil Erosion Management Research in Asian Catchments: Methodological Approaches and Initial Results, edited by A. R. Maglinao and R. N. Leslie. Colombo, Sri Lanka, International Water Management Institute. pp 85-93.

De Voogt, K., Kite, G., Droogers, P. & Murray-Rust, H. 2000, Modeling Water Allocation between Wetlands and irrigated Agriculture: Case Study of the Gediz Basin, Turkey, IWMI Working Paper 1. Colombo, Sri Lanka, International Water Management Institute.

de Zulueta, J., Mujtaba, S.M. & Shah, I.H. 1980, Malaria control and long term periodicity of the disease in Pakistan. Transactions of the Royal Society of Tropical Medicine and Hygiene 74, 624-632.

Dhawan, B. D. 1982, Development of Tubewell Irrigation in India. New Dehi, Agricole Publishing Academy.

Dhawan, B. D. 1985, Output impact according to main irrigation sources: Empirical evidence from four selected states. Presented at the INSA National Seminar on 'Water Management: Key to Development of Agriculture', New Delhi, 28-30 April 1986.

Dinar, A. & Saleth, R.M. 2005, Issues in water pricing reforms: From getting correct prices to setting appropriate institutions. In: *International Yearbook of Environmental and Resource Economics*, edited by Tom Tietenberg and Henk Folmer. New Horizons in Environmental Economics Series, Volume IX. Cheltenham, Edward Elgar Publishing Pvt. Ltd., UK, pp. 1-51.

Donnelly, M.J., Birley, M.H. & Konradsen, F. 1997a, An Investigation of the Relationship between Depth to Groundwater and Malaria Prevalence, Punjab, Pakistan. IIMI Working Paper No. 40. Colombo, Sri Lanka, International Irrigation Management Institute.

Donnelly, M.J., Konradsen, F. & Birley, M.H. 1997b, Malaria treatment-seeking behaviour in the southern Punjab. Annals of Tropical Medicine and Parasitology 91, 665-667.

Dowling, T. (ed.). 1996, Sustainable Land Management in the South Pacific (IBSRAM/PACIFICLAND). Network Document 19. Bangkok, Thailand, International Board for Soil Research and Management.

Drechsel, P. 2000, Lessons from on-farm research on sustainable land management in sub-Saharan Africa – Results from an international meeting in Abengourou, Cote d'Ivoire. In: *Farmers and Scientists in a Changing Environment: Assessing Research in West Africa*, edited by G. Renard, S. Krieg, P. Lawrence and M. Margraf Van Oppen. Weikersheim, Germany, Verlag. page 319-326.

Drechsel, P. & Gyiele, L. 1998, On Farm Research on Sustainable Land Management in Subsaharan Africa: Approaches, Experiences and Lessons. IBSRAM Proceedings 19. Bangkok, Thailand, International Board for Soil Research and Management. 254 pp.

Drechsel, P. & Gyiele, L. 2000, Methods for the Economic Assessment of the On- and Off Site Impacts of Nutrient Mining. Issues in Sustainable Land Management 7. Bangkok, Thailand, International Board for Soil Research and Management. 50 pp.

Drechsel, P., Giordano, M. & Gyiele, L. 2004, Valuing Nutrients in Soil and Water: Concepts and Techniques with Examples from IWMI Studies in the Developing World. IWMI Research Report No. 82. Colombo, Sri Lanka, International Water Management Institute.

Drechsel, P., Kunze, D. & Penning de Vries, F.W.T. 2001, Soil nutrient depletion and population growth in Sub-Saharan Africa: A Malthusian nexus? Population and Environment 22 (4), 411-423.

Droogers, P. & Kite, G. 2001a, Estimating Productivity of Water at Different Spatial Scales Using Simulation Modeling. IWMI Research Report 53. Colombo, Sri Lanka, International Water Management Institute.

Droogers, P. & Kite, G. (eds.) 2001b, Special issue of Journal of Hydrology on Assessing Actual Evapotranspiration.

Droogers, P., Seckler, D. & Makin, I. 2001, Estimating the Potential of Rain-fed Agriculture. IWMI Working Paper 20. Colombo, Sri Lanka, International Water Management Institute.

Droogers, P., Bastiaanssen, W.G.M., Gieske, A., Toomanian, N. & Akbari, M. 2001, Assessment of Irrigation Performance Using NOAA Satellite Imagery. Iran-IWMI Research Report 6. Colombo, Sri Lanka, International Water Management Institute.

Dubash, N. K. 2002, Tubewell Capitalism: Groundwater Development and Agrarian Change in Gujarat. New Delhi, India, Oxford University Press, xiii, 287p.

Dumanski, J. & Craswell, E. T. 1998, Resource management domains for evaluation and management of agro-ecological systems. In: International Workshop on Resource Management Domains, 1-13. Bangkok, Thailand, International Board for Soil Research and Management.

Dupin, B., Phantahvong, K.B., Chanthavongsa, A. & Valentin, C. 2002, Assessment of tillage erosion rates on steep slopes in the northern Lao PDR. The Lao Journal of Agriculture and Forestry 4, 52-59.

Economist. 2004, Putting the World to Rights. June 3, 2004.

Eddleston, M., Karalliedde, L., Piola, J.C., Smith, T., Buckley, N., Isbister, G., Singh, S., Fernando, R., Wilkinson, R., Konradsen, F., Senanayake, N., Smit, L. &

Kappagoda, S. 2002, Pesticide poisoning in the developing world – a minimum pesticides list. Lancet 360, 1163-1167.

Ekanayake, L. & van der Hoek, W. 2002, Dental caries and developmental defects of enamel in relation to fluoride levels in drinking water in an arid area of Sri Lanka. Caries Research 36, 398-404.

Ekanayake, L. & van der Hoek, W. 2003, Prevalence and distribution of enamel defects and dental caries in a region with different concentrations of fluoride in drinking water in Sri Lanka. International Dental Journal 53,243-248.

Elkaduwa, W. K. B. & Sakthivadivel, R. 1999, Use of Historical Data as a Decision Support Tool in Watershed Management: A Case Study of the Upper Nilwala Basin in Sri Lanka. IIMI Research Report 26. Colombo, Sri Lanka, International Water Management Institute.

Ensink, J. IWMI, personal communication.

Ensink, J. H. J., Jensen, P. K., Aslam, M. R. & van der Hoek, W. 2000, Integration of irrigation and rural drinking water. 26th WEDC Conference, Water, Sanitation & Hygiene: Challenges of the Millennium, 5-9 November, Dhaka, Bangladesh. Water Provision, pp.329-332.

Ensink, J.H.J., Aslam, M.R., Konradsen, F., Jensen, P.K. & Van der Hoek, W. 2002b, Linkages between Irrigation and Drinking Water in Pakistan. Working Paper No. 46. Colombo, Sri Lanka, International Water Management Institute.

Ensink, J.H.J., Mahmood, T., van der Hoek, W., Raschid-Sally, L. & Amerasinghe, F.P. 2003, A nation-wide assessment of wastewater use in Pakistan: an obscure activity or a vitally important one? Water Policy (in press).

Ensink, J.H., van der Hoek, W., Matsuno, Y., Munir, S. & Aslam, M.R. 2002a, Use of Untreated Wastewater in Peril-Urban Agriculture in Pakistan: Risks and Opportunities. Research Report No. 64. Colombo, Sri Lanka, International Water Management Institute.

Enters, T. 1998, Methods for the Economic Assessment of the On- and Off Site Impacts of Soil Erosion. Issues in Sustainable Land Management 2. Bangkok, International Board for Soil Research and Management. 60 pp.

Eusof, Z., Phien, T., Santoso, D., Boonchee, S., Sengtaheunghoung, O., Dixin, Y. & Amada, A.B. 2002, Residue management in alley cropping systems for erosion control and soil improvement on sloping lands in Asia. In: Management of Sloping Lands for Sustainable Agriculture. Final Report of ASIALAND Sloping Lands Project, Phase 4, edited by F. Penning de Vries. Bangkok, International Water Management Institute.

Faerge, J., Magid, J. & Penning de Vries, F.W.T. 2001, Urban nutrient balance for FMIS, 1991, Farmer Managed Irrigation Systems Newsletter July 1991. Colombo, Sri Lanka: International Irrigation Management Institute.

Falkenmark, M. 1986, Freshwater-Time for a modified approach. Ambio 15 (4),192-200.

FAO (Food and Agriculture Organization of the United Nations). 2000, The State of World Fisheries and Aquaculture. Rome, Food and Agriculture Organization of the United Nations.

FAO. 2002, World agriculture: Towards 2015/2030 – Summary report. Rome, Italy: FAO. x, 97p.

FAO. 2003, Groundwater Management: The Search for Practical Approaches. Water Reports 25. Rome, FAO.

FAO. 2003, The Irrigation Challenge. Rome, IPTRID Secretariat, FAO.

Feenstra, S., Hussain, R. & van der Hoek, W. 2000a, Prevalence of Intestinal Parasites in the Southern Punjab, Pakistan. IWMI Pakistan Report No. 101. Lahore, Pakistan, International Water Management Institute. 27 pp.

Feenstra, S., Hussain, R. & van der Hoek, W. 2000b, Health Risks of Irrigation with Untreated Urban Wastewater in the Southern Punjab, Pakistan. IWMI Pakistan Report No. 107. Lahore, Pakistan, International Water Management Institute.

Feenstra, S., Jabbar, A., Masih, R. & Jehangir, W.A. 2000c, Health Hazards of Pesticides in Pakistan. IWMI Pakistan Report No.100. International Water Management Institute and Pakistan Agricultural Research Council, Colombo/Islamabad.

Fietelson, E. 2003, Sustainable groundwater management: Has regulation worked in Israel the Mecca of water management? In: Groundwater *Governance in Asia: The Challenge of Taming a Colossal Anarchy,* edited by IWMI-Tata Water Policy Program, Anand.

Forget, G. & Lebel, J. 2001, An ecosystem approach to human health. Int. J. Occup. Environ. Health. 7, Suppl, S3-S36.

Foster, S., Chilton, J., Moench, M., Cardy, F. & Schiffler, M. 2000, Groundwater in Rural Development: Facing the Challenges of Supply and Resource Sustainability. World Bank Technical Paper no.463. Washington, D.C., USA, World Bank.xvi, 98p.

Frederiksen, H. D. & Vissia, R. J. 1998, Considerations in Formulating the Transfer of Services in the Water Sector. Colombo, Sri Lanka, International Water Management Institute.

Galbraith, H., Amerasinghe, P. & Huber-Lee, A. 2005, The Effects of Agricultural Irrigation on Wetland Ecosystems in Developing Countries – A Literature Review. Comprehensive Assessment Discussion Paper 1. Colombo, Sri Lanka, Comprehensive Assessment Secretariet.

Gallup, J.L. & Sachs, J.D. 2001, The economic burden of malaria. American Journal of Tropical Medicine and Hygiene 64, 85-96.

Gardiner, C., Biggar, R.J., Collins, W.E. & Nkrumah, F.K. 1984, Malaria in urban and rural areas of southern Ghana: A survey of parasiteamia, antibodies, and antimalarial practices. Bulletin of the World Health Organization 62, 607-613.

Geertz, C. 1963, Agricultural Involution. Berkeley, University of California,

Giordano, M. 2006, Agricultural groundwater use and rural livelihoods in Sub-Saharan Africa: A first-cut assessment. Hydrogeology Journal. (In press)

Giordano, M. A. 2003, "Assessing IWMI's Research Impacts: A Framework for Action". In: *Methods for Assessing the Impacts of Natural Resource Management Research: A Summary of the Proceedings of the ICRISAT-NCAP/ICAR International Workshop,* edited by B. Shiferaw and H.A. Freeman. Patancheru, India: ICRISAT.

Giordano, M., Samad, M. & Namara, R. Forthcoming. Assessing the Outcomes of IWMI's Research and Interventions on Irrigation Management Transfer (IMT). Contributing chapter to the SPIA Commissioned book on CGIAR NRM Impact Assessment.

Gleick, P. 2003, Soft path's solution to 21st-century water needs. Science 320, 1524–1528.

Gleick, P.H. 1998, The human right to water. Water Policy 1, 487-503.

GoI (Government of India). 1999, Integrated Water Resources Development: A Plan for Action, Report of the National Commission for Integrated Water Resources Development Plan, Vol. I. New Delhi, Ministry of Water Resources, Sept.

Goodfellow, A., Lewis, D., Madramootoo, C. & Boelee, E. 2003, Entomological assessment of malaria vector populations in two neighboring regions of southern Sri Lanka. Poster presented at IDRC Ecohealth Forum, Montreal, Canada. May 2003.

Greenland, D.J., Bowen, G., Eswaran, H., Rhoades, R. & Valentin, C. 1994, Soil, Water, and Nutrient Management: A New Agenda. IBSRAM Position Paper. Bangkok, Thailand, International Board for Soil Research and Management. 72 pp.

Guerra, L. C., Bhuiyan, S.I., Tuong, T.P. & Barker, R. 1998, Producing More Rice with Less Water from Irrigated Systems. IWMI SWIM Paper 5. Colombo, Sri Lanka. International Water Management Institute.

GWP (Global Water Partnership). 2000, Towards Water Security: A Framework for Action. GWP, Stockholm, Sweden.

Hagino, N. 1968, Study of Itai-Itai disease. 13th annual meeting of the Toyama medical Society. Cited in: Heavy Metal Pollution in Soils of Japan, edited by K. Kitagishi and I. Yamane. Japan Science Society Press, Tokyo.

Harmancioglu, N., Alpaslan, N. & Boelee, E. 2001, Irrigation, Health and Environment: A Review of Literature from Turkey. Working Paper No. 6. Colombo, Sri Lanka, International Water Management Institute.

Hassan, M. ul, Noble, A. & Kazbekov, J. 2004, "Bright Spots" in Uzbekistan, Reversing Land and Water Degradation while Improving Livelihoods: Key Development and Sustaining Ingredients for Slow Transition Economies of the Former Soviet Union. IWMI Research Report 78. Colombo, Sri Lanka, International Water Management Institute. vi, 47p.

Hazell, P. & Johnson, M. 2002, Cutting Hunger in Africa through Smallholder-led Agricultural Growth, IFPRI Technical Paper in Support of USAID's Initiative to End Hunger in Africa (IEHA). World Summit on Sustainable Development (WSSD).

Heath, R. 1987, Basic Groundwater Hydrology. Denver, USA, US Geological Survey.

Hekmat, A. 2002, Overexploitation of groundwater in Iran: Need for an integrated water policy. Paper for the IWMI-ICAR-Colombo Plan-sponsored Policy Dialogue on "Forward-Thinking Policies for Groundwater Management: Energy, Water Resources, and Economic Approaches' organized at India International Center, New Delhi, 2-6 September 2002.

Herrel, N., Amerasinghe, F.P., Ensink, J., Mukhtar, M., van der Hoek, W. & Konradsen, F. 2001, Breeding of Anopheles mosquitoes in irrigated areas of South Punjab, Pakistan. Medical and Veterinary Entomology 15,236-248

Herrel, N., Amerasinghe, F.P., Ensink, J. Mukhtar, M., van der Hoek, W. & Konradsen, F. 2004, Adult anopheline ecology and malaria transmission in an irrigated area of south Punjab, Pakistan. Medical & Veterinary Entomology 18, 1-12.

Hinrichsen, D., Robey, B. & Upadhyay, U.D. 1997, Solutions for a Water-Short World. Population Reports, Series M, No. 14. Baltimore, Johns Hopkins School of Public Health, Population Information Program, December 1997.

Hoeksta, A.Y. & Hung, P.Q. 2003, Virtual Water Trade: A Quantification of Virtual Water Flows between Nations in Relation to International Crop Trade. Value of Water Research Report Series No.11. Delft, the Netherlands, IHE.

Hong, L, Li, Y.H., Deng, L. Chen, C.D., Dawe, D. & Barker, R. 2001, Analysis of changes in water allocations and crop production, Zanghe Irrigation System and District – 1966 to 1998. In: Water Saving Irrigation for Rice, edited by R. Barker, R. Loeve, Y.H. Li and T.P.Tuong. Colombo, Sri Lanka: International Water Management Institute.

Huber-Lee, A., Yates, D., Purkey, D., Yu, W. & Runkle, B. 2003, Water, Climate, Food, and Environment in the Sacramento Basin. ADAPT Project Final Report. P. 59. Netherlands, Wageningen Agricultural University.

Hunter, J.M., Rey, L. Chu, K. Y., Adekolu-John, E.O. & Mott, K.E. 1993, Parasitic Diseases in Water Resources Development. WHO, Geneva. 152 pp.

Hussain, I. 2005, Pro-poor Intervention Strategies in Irrigated Agriculture in Asia, Final Synthesis Report. Colombo Sri Lanka, International Water Management Institute.

Hussain, I. & Hanjra, M. 2003, Does irrigation water matter for rural poverty alleviation? Water Policy 5 (5/6), 429-442.

Hussain, I., Marikar, F., & Jehangir, W. 2000, Productivity and Performance of Irrigated Wheat Farms across Canal Commands in the Lower Indus Basin. IWMI Research Report 44. Colombo, Sri Lanka, International Water Management Institute.

Hussain, I., Raschid, L. Hanjra, M.A., Marikar, F. & Van der Hoek, W. 2001, A Framework for Analyzing Socioeconomic, Health and Environmental Impacts of Wastewater Use in Agriculture in Developing Countries. IWMI Working Paper No. 26. Colombo, Sri Lanka, International Water Management Institute.

Hussain, I., Raschid, L., Hanjra, M.A., Marikar, F. & Van der Hoek, W. 2002, Wastewater Use in Agriculture: Review of Impacts and Methodological Issues in Valuing Impacts. IWMI Working Paper No. 37. Colombo, Sri Lanka, International Water Management Institute.

Hussain, I., Sakthivadivel, R., Amarasinghe, U., Mudasser, M. & Molden, D. 2003, Land and Water Productivity of Wheat in the Western Indo-Gangetic Plains of India and Pakistan: A Comparative Analysis. Research Report 65. Colombo, Sri Research Report 22. Colombo, Sri Lanka. International Water Management Institute.

IAC (Inter Academy Council.) 2004, Realizing the promise and potential of African Agriculture. (http://www.interacademycouncil.net/report.asp?id=6959)

IBSRAM (International Board for Soil Research and Management). 1992, ASIALAND: The Management of Sloping Lands for Sustainable Agriculture in Asia (1989-91). Bangkok, Thailand.

IBSRAM. 1995, ASIALAND: The Management of Sloping Lands for Sustainable Agriculture in Asia (1992-95). Bangkok, Thailand.

IBSRAM. 1997, Towards Sustainable Land Management in the 21st Century: The IBSRAM Vision. Bangkok, Thailand.

IBSRAM. 1998, ASIALAND: The Management of Sloping Lands for Sustainable Agriculture in Asia (1995-97). Bangkok, Thailand.

IBSRAM. 1999, Common Ground: Medium Term Plan. Bangkok, Thailand.

IDRC (International Development Research Centre). 2005, Brochure on Outcome Mapping: The Challenges of Assessing Development Impacts. Available on the International Development Research Centre website at (http://web.idrc.ca/en/ev-64698-201-1-DO_TOPIC.html)

IFAD (International Fund for Agricultural Development). 2000, Rural poverty report, 2001: The challenge of ending rural poverty. Rome, Oxford Press, IFAD.

Ijumba, J. N. & Lindsay, W. 2001, Impact of irrigation on malaria in Africa: paddies paradox. Medical and Veterinary Entomology 15, 1-11.

Ilao, R.O., Duque, C.M., Tiongco, L.E., Villano, M.G., Carpina, N.V., Santos, B.G., De Guzman, M.T.L., Quita, R.S., Yadao, R.S. & Mugot, I.O. 2002, Management of Soil Erosion Consortium (MSEC): An Innovative Approach to Sustainable Land Management in the Philippines. Project Completion Report.

Ines, A.V.M., Droogers, P., Makin, I. W. & Gupta, A. D. 2001, Crop Growth and Soil Water Balance Modeling to Explore Water Management Options. IWMI Working Paper 22. Colombo, Sri Lanka, International Water Management Institute.

Inocencio, A., Merrey, J. D. & Sally, H. 2003, Innovative Approaches to Agricultural Water Use for Improving Food Security in Sub-Saharan Africa. IWMI Working Paper 55. Colombo, Sri Lanka, International Water Management Institute.

Inocencio, A., Kikuchi, M., Merrey, D., Tonosaki, M., Maruyama, A., Jong, I. de, Sally, H. & Penning de Vries, F. 2005, Lessons from Irrigation Investment Experiences: Cost-reducing and Performance-enhancing Options for Sub-Saharan Africa, Final Report, African Development Bank – IWMI Collaborative Program on Investments in Agricultural Water Management in Sub Saharan Africa. Colombo, Sri Lanka, International Water Management Institute.

Inthansothi, S., Jirasukteveekul, W., Adirektrakarn, W., Ratcahdawong, S. & Boonsaner, A. 2000, Catchment approach to combating soil erosion in Thailand. In: Soil erosion management research in Asian catchments: Methodological approaches and initial results. Proceedings of the 5th Management of Soil Erosion Consortium (MSEC) Assembly, edited by A. R. Maglinao and R. N. Leslie. Bangkok, Thailand, IWMI Southeast Asia Regional Office, Thailand. pp. 233-252.

IWMI (International Water Management Institute), 2000, Water Issues for 2025: A Research Perspective. Colombo, Sri Lanka, International Water Management Institute

IWMI. 2000, World's Water Supply and Demand 1995-2025. Colombo, Sri Lanka, IWMI.

IWMI. 2001a, Strategic Plan 2000-2005. Colombo, Sri Lanka.

IWMI. 2001b, Research Program on Institutional Support Systems for Sustainable Local Management of Irrigation in Water-Short Basins, Final Report. Submitted to the BMZ and GTZ, Federal Republic of Germany.

IWMI. 2003, Growing More Rice with Less Water: Increasing Water Productivity in Rice-Based Cropping Systems. Progress of Research 1 July 2002 to 30 June 2002. IWMI Working Paper 54. Colombo, Sri Lanka.

IWMI. 2003a, Innovations in Groundwater Recharge. Water Policy Briefing No. 1. Colombo, Sri Lanka, International Water Management Institute.

IWMI. 2003b, Groundwater Governance in Asia: The Challenge of Taming a Colossal Anarchy, IWMI Contribution to III World Water Forum, Kyoto for the Session on "Water, Food and Environment". Anand, IWMI-Tata Water Policy Program.

IWMI. 2004, *Water for the Future – Linking Irrigation and Water Allocation in the Zayandeh Rud Basin, Iran,* edited by H. Murray-Rust and P. Droogers. IWMI Monograph. Colombo, Sri Lanka.

IWMI and GDRS (General Directorate of Rural Services), Turkey. 2000, Irrigation in the Basin Context: The Gediz Study. Colombo, Sri Lanka, International Water Management Institute.

Janakarajan, S. 1994, Trading in groundwater: A source of power and accumulation, In: *Selling Water: Conceptual and Policy Debates over Groundwater Markets in India,* edited by Marcus Moench. Ahmedabad, India, VIKSAT.

Janeau, J. L., Bricquet, J.P., Planchon, O. & Valentin, C. 2003a, Soil crusting and infiltration on very steep slopes in northern Thailand. The European Journal of Science 54, 543:553.

Janeau, J.L., Maglinao, A.R., Loent, C., Bricquet, J.P. & Boonsaner, A. 2003b, Off-site effect of soil erosion: A case study of the Mae Thang reservoir in northern Thailand.

Paper presented at the KU-IWMI Seminar on Scientific Cooperation, 26 March 2003. Bangkok, Thailand, Kasetsart University.

Javier, E. & Voss, J. 2003, Towards Integrated Natural Resources Management Research. in Research Towards Integrated Natural Resources Management. CGIAR. pp.3-18.

Jayakody, D.S. 1993, Causes for the decline of finfish and shellfish catches of three selected lagoons: Malala, Rekawa and Mawella on the south coast of Sri Lanka and the current status. Pp. 11-13. In: Proceedings of the First Annual Sessions of the National Aquatic Resources Agency (NARA). 2 November 1993. Colombo, Sri Lanka.

Jehangir, W. A. & Horinkova, V. 2002, Institutional Constraints to Conjunctive Water Management in the Rechna Doab. Working Paper No. 50. Lahore, Pakistan, International Water Management Institute.

Jehangir, W. A., Qureshi, A. S., & Ali, N. 2002, Conjunctive Water Management in the Rechna Doab: An Overview of Resources and Issues. IWMI Working Paper 48. Lahore, Pakistan, International Water Management Institute, vi. 55p.

Jensen, P.K., Aalbæk, B., Aslam, R. & Dalsgaard, A. 2001, Specificity for field enumeration of *Eschericha coli* in tropical surface waters. Journal of Microbiological Methods, 45, 135-141.

Jensen, P.K., Ensink, J., Jayasinghe, G., van der Hoek, W., Cairncross, S. & Dalsgaard, A. 2002, Domestic transmission routes of pathogens: The problem of in-house contamination of drinking water during storage in developing countries. Tropical Medicine and International Health, 7(7), 604-609.

Jensen, P.K., Ensink, J.H.J., Jayasinghe, G., van der Hoek, W., Cairncross, S. & Dalsgaard, A. 2003, Effect of chlorination of drinking-water on water quality and childhood diarrhoea in a village in Pakistan. Journal of Health, Population and Nutrition 21, 26-31.

Jobin, W. 1999, Dams and Disease: Ecological Design and Health Impacts of Large Dams, Canals and Irrigation Systems. E & FN Spon (Taylor & Francis Group). 580 pp.

Johnson, S.H. III. 1997, Irrigation Management Transfer in Mexico: A Strategy to Achieve Irrigation District Sustainability. IWMI Research Report 16. Colombo, Sri Lanka, International Water Management Institute.

Joy, K J. & Paranjape, S. 2003, Tail-enders and other deprived in canal irrigation systems: Maharashtra, Paper for The National Workshop on "Tail-enders and Other Deprived in Canal Irrigation Systems", organized by Development Support Centre at Ahmedabad during November 28-29, 2003.

Kahnert, F. & Levine, G. (eds.), 1993. Groundwater Irrigation and the Rural Poor: Options for Development in the Gangetic Basin. Washington, D.C., USA, World Bank. A World Bank Symposium, April 1989. ix, 222p.

Karlberg, L. & Penning de Vries, F.W.T. 2004, Exploring potentials and constraints of low-cost drip irrigation with saline water in sub-Saharan Africa. Physics and Chemistry of the Earth 29:1035-1042.

Keiser, J., Utzinger, J. & Singer, B.H. 2002, The potential of intermittent irrigation for increasing rice yields, lowering water consumption, reducing methane emissions, and controlling malaria in African rice fields. Journal of the American Mosquito Control Association 18, 329-340.

Keller, J., Keller, A., & Davids, G. 1998, River basin development phases and implications of closure. Journal of Applied Irrigation Science 33(2),145-163.

Keller, A., Saktivadivel, R. & Seckler, D. 2000, Water Scarcity and the Role of Storage in Development. Research Report 39. Colombo, Sri Lanka, International Water Management Institute. 28 pp.

Keraita, B., Drechsel, P., Huibers, F. & Raschid-Sally, L. 2002, Wastewater use in informal irrigation in urban and peril-urban areas of Kumasi, Ghana. Urban Agriculture Magazine 8, 11-13.

Khan, S., Rana, T., Ullah, E. W., Christen, E. W. & Ahmed, H. M. N. 2003, Investigating Conjunctive Water Management Options Using a Dynamic Surface-Groundwater Modelling Approach: A Case Study of Rechna Doab. CSIRO Technical Report, 35/03, June 2003.

Kido, T., Honda, R., Tsuritani, I., Yamaya, H., Ishizaki, M., Yamada, Y. & Nogawa, K. 1988, Progress of renal dysfunction in inhabitants environmentally exposed to cadmium. Archives of Environmental Health 43, 213-217.

Kido, T., Nogawa, K., Honda, R., Tsuritani, I.., Ishizaki, M., Yamada, Y. & Nakagawa, H. 1990, The association between renal dysfunction and osteopenia in environmental cadmium-exposed subjects. Environmental Research 51, 71-82.

Kijne, J.W. & Kuper, M. 1995, Salinity and sodicity in Pakistan's Punjab: A threat to sustainability of irrigated agriculture. Water Resources Development 11(1), 73-86

Kijne, J. W. & Kuper, M. 1996, Sustainable management of land and water resources in irrigated semi-arid and tropical regions. In: Soil Quality and Agricultural Sustainability, edited by R. Lal. Chelsea, MI, USA, Ann Arbor Press. pp.67-85.

Kijne, J. W., Barker, R. & Molden, D. (eds.) 2003, Water Productivity in Agriculture: Limits and Opportunities for Improvement. Comprehensive Assessment of Water Management in Agriculture, Volume 1. Wallingford, UK, CABI.

Kijne, J., Prathapar, S.A., Wopereis, M.C.S. & Sahrawat, K.L. 1998, How to Manage Salinity in Irrigated Lands: A Selective Review with Particular Reference to Irrigation in Developing Countries. IWMI SWIM Paper 2. Colombo, Sri Lanka, International Water Management Institute.

Kikuchi, M., Barker, R., Weligamage, P. & Samad, M. 2002, Irrigation Sector in Sri Lanka: Recent Investment Trends and the Development Path Ahead. Research Report 62. Colombo, Sri Lanka, International Water Management Institute.

Kikuchi, M., Weligamage, P., Barker, R., Samad, M., Kono, H. & Somaratne, H.M. 2003, Agro-well and Pump Diffusion in the Dry Zone of Sri Lanka: Past Trends, Present Status, and Future Prospects. Research Report 66. Colombo, Sri Lanka, International Water Management Institute.

Kite, G. 2000, Developing a Hydrological Model for the Mekong Basin: Impacts of Basin Development on Fisheries Productivity. IWMI Working Paper 2. Colombo, Sri Lanka, International Water Management Institute.

Kite, G. & Droogers, P. 2000a, Integrated Basin Modeling. IWMI Research Report 43. Colombo, Sri Lanka, International Water Management Institute.

Kite, G. & Droogers, P. 2000b, Comparing Estimates of Actual Evapotranspiration from Satellites, Hydrological Models and Field Data: A Case Study from Western Turkey. IWMI Research Report 42. Colombo, Sri Lanka, International Water Management Institute.

Kite, G., Droogers, P., Murray-Rust, H. & de Voogt, K. 2001, Modeling Scenarios for Water Allocation in the Gediz Basin, Turkey. IWMI Research Report 50. Colombo, Sri Lanka, International Water Management Institute.

Klare, M.T. 2001, The new geography of conflict. Foreign Affairs 80 (4), 49-61.

Kleinschmidt, I., Omumbo, J., Briet, O., van de Giesen, N., Sogoba, N., Kamasenu, N., Windmeijer, P., Moussa, M. & Teuscher, T. 2001, An empirical malaria distribution map for West Africa. Tropical Medicine and International Health 6, 779-786.

Klinkenberg, E., Konradsen, F., Herrel, N., Mukhtar, M., van der Hoek, W., Amerasinghe, W.P. 2004, Malaria vectors in the changing environment of the Southern Punjab, Pakistan. Transactions of the Royal Society of Tropical Medicine and Hygiene (in press).

Klinkenberg, E., McCall, P.J., Wilson, M.D., Akoto, A. & Donnelly, M.J. 2003b, Impact of urban agriculture on malaria prevalence in young children in two cities of Ghana. Proceedings of the IDRC Forum on Ecosystem Approaches to Human Health, Montreal, Canada. 18-23 May 2003. (Abstract).

Klinkenberg, E., van der Hoek, W., Amerasinghe, F.P., Gunawardena, D.M., Muttawatte, L. & Jayasinghe, G. 2003a, Malaria and Agriculture: A Risk Analysis in Southern Sri Lanka. Research Report No. 68. Colombo, Sri Lanka, International Water Management Institute.

Kloezen, H. W. & Garcés-Restrepo, C. 1998, Assessing Irrigation Performance with Comparative Indicators: The Case of the Alto Rio Lerma Irrigation District, Mexico. Colombo, Sri Lanka, International Water Management Institute.

Kloezen, H.W., Garcés-Restrepo, C. & Johnson III, S. H. 1997, Impact Assessment of Irrigation Management Transfer in the Alto River Lerma Irrigation District, Mexico. IWMI Research Report 15. Colombo, Sri Lanka, International Water Management Institute.

Kolavalli, S. & David, L. C. 1989, Groundwater markets in Gujarat, India. Water Resources Development 5 (1).

Kolsky, P.J. 1993, Diarrheal disease: Current concepts and future challenges; water, sanitation and diarrhea: the limits of understanding. Transactions of the Royal Society of Tropical Medicine and Hygiene 87, 43-46.

Konboon, Y., Wijnhoud, J.D. & Lefroy, R.D.B. 2001, Nutrient Management in Rain-fed Lowland Rice-Based Systems in Northeast Thailand. Proceedings Workshop Nutrient Recycling. Bangkok, Thailand, Kasetsart University.

Konradsen, F. & van der Hoek, W. (eds.). 1998, Health and Irrigation. Proceedings of the Copenhagen workshop on health and irrigation. Ministry of Foreign Affairs, Copenhagen, Denmark, 18-20 August 1997. Colombo, Sri Lanka, International Irrigation Management Institute. 32 pp.

Konradsen, F., Amerasinghe, F.P., van der Hoek, W. & Amerasinghe, P.H. 2000a, Malaria in Sri Lanka: Current Knowledge on Transmission and Control. Colombo, Sri Lanka, International Water Management Institute. 77 pp.

Konradsen, F., Amerasinghe, P.H., Perera, D., van der Hoek, W. & Amerasinghe, F.P. 2000b, A village treatment center for malaria: Community response in Sri Lanka. Social Science & Medicine 50, 879-889.

Konradsen, F., Amerasinghe, P., van der Hoek, W., Amerasinghe, F., Perera, D. & Piyaratne, M. 2003a, Strong association between house characteristics and malaria vectors in Sri Lanka. American Journal of Tropical Medicine and Hygiene 68 (2), 177-181.

Konradsen, F., Matsuno, Y., Amerasinghe, F.P., Amerasinghe, P.H. & van der Hoek, W. 1998b, *Anopheles culicifacies* breeding in Sri Lanka and options for control through water management. Acta Tropica 71, 131-138.

Konradsen, F., Steele, P., van der Hoek, W., Amerasinghe, P.H. & Amerasinghe, F.P. 1999, Cost of malaria control in Sri Lanka. Bulletin of the World Health Organization 77, 301-309.

Konradsen, F., Stobberup, K. A., Sharma, S.K., Gulati, O.T. & van der Hoek, W. 1998a, Irrigation water releases and *Anopheles culicifacies* abundance in Gujarat, India. Acta Tropica 71, 195-197.

Konradsen F., van der Hoek, W., Amerasinghe, P.H. & Amerasinghe, F.P. 1997a, Measuring the economic cost of malaria to households in Sri Lanka. American Journal of Tropical Medicine and Hygiene 56, 656-660.

Konradsen, F., van der Hoek, W., Amerasinghe, P.H., Amerasinghe, F.P. & Fonseka, K.T. 1997b, Household responses to malaria and their costs: A study from rural Sri Lanka. Transactions of the Royal Society of Tropical Medicine and Hygiene 91, 127-130.

Konradsen, F., van der Hoek, W. Cole, D.C., Hutchinson, G., Daisley, H., Singh, S. & Eddleston, M. 2003b, Reducing acute poisoning in developing countries – options for restricting the availability of pesticides. Toxicology 192, 249-261.

Konradsen, F., van der Hoek, W. Perry, C. & Renault, D. 1997c, Water: Where from, and for whom? World Health Forum 18(1), 41-3.

Koppen, B. van. 2002, A Gender Performance Indicator for Irrigation. Concepts, Tools, and Applications. IWMI Research Report 59. Colombo, Sri Lanka, International Water Management Institute. 60 pp.

Krishnasamy, S., Amerasinghe, F.P., Sakthivadivel, R., Ravi, G., Tewari, S.C. & van der Hoek, W. 2003, Strategies for Conserving Water and Effecting Mosquito Vector Control in Rice Ecosystems: A Case Study from Tamil Nadu, India. Working Paper No. 56. Colombo, Sri Lanka, International Water Management Institute.

Kularatne, M. G. 1999, Fishermen without fish. The effects of productivity decline in lagoon fisheries on a fishing and farming community and its use of natural resources. A case study of Malala lagoon, Hambantota, Sri Lanka. M.Sc. thesis. Enschede, The Netherlands, International Institute for Aerospace Survey and Earth Sciences.

Kumar, A. 1997, Urban malaria and its control in India. Journal of Parasitic Diseases 21, 83-88.

Kumar, M. D. 2005, Impact of electricity prices and volumetric water allocation on energy and groundwater demand management: Analysis from Western India. Energy Policy 33, (2005):39-51.

Kumar, M.D. & Singh, O.P. 2005, Which Water Counts? Blue and Green Water Use and Productivity in the Narmada Basin. IWMI-TATA Water Policy Research Highlights No. 8, 2005.

Kumar, M. D., Singh, O. P. & Singh, K. 2002, Groundwater Depletion and Its Socio-ecology Consequences in Sabarmati River Basin. *Proceedings of Symposium on Intensive Use of Groundwater,* Valencia, Spain, 10-14 December, 2002.

Kumar, M. D., Shah, T., Bhatt, M. & Kapadia, M. 2003, Dripping water to a water guzzler? A techno-economic evaluation of the efficiency of drip irrigation in Alfalfa. Paper presented at the IWMI Tata Annual Partners' Meet, Anand, 17-19 February 2004.

Lacey, L.A. & Lacey, C.M. 1990, The medical importance of riceland mosquitoes and their control using alternatives to chemical insecticides. Journal of the American Mosquito Control Association 6 (Suppl.), 1-93.

Lacroix, M., Kite, G. & Droogers, P. 2000, Using Datasets from the Internet for Hydrological Modeling: An Example from the Kucuk Menderes Basin, Turkey. IWMI Research Report 40. Colombo, Sri Lanka, International Water Management Institute.

Lal, R. 1995, Erosion-crop productivity relationships for soils of Africa. Soil Sci. Soc. Am. 59, 661-667.

Lashari, B. & Murray-Rust, H. 2000, Remodeling of Outlets in Three Pilot Distributaries: The Farmer-Managed Irrigation Project in Sindh Province Pakistan. IWMI Working Paper 13. Colombo, Sri Lanka, International Water Management Institute.

Lashari, B., Murray-Rust, H. & Talpur, M. 2000, Achieving Stable Canal Conditions Following Remodeling: The Case Study of Bareji Distributary Mirpurkhas. IWMI Working Paper 12. Colombo, Sri Lanka, International Water Management Institute

Laxminarayan, R. 2004, Does malaria improve households' living standards? Tropical Medicine and International Health (in press). Citation based on paper at (http://www2.ncsu.edu/unity/lockers/users/v/vksmith/ opportunites/ malaria_viet.pdf)

Lefroy, R.D.B., Bechstedt, H.D. & Rais, M. 2000, Indicators for sustainable land management based on farmer surveys in Vietnam, Indonesia, and Thailand. Agriculture, Ecosystems and Environment 81, 137-146.

Lemly, A.D., Kingsford, R.T., & Thompson, J. R. 2000, Irrigated agriculture and wildlife conservation: Conflict on a global scale. *Environmental Management* 25, 485-512.

Levine, G., Galvan, C. C., Garcés-Restrepo, C. & Johnson, S. III. 1998, Performance of Two Modules in the Lagunera Region of Mexico. IWMI Research Report 23. Colombo, Sri Lanka, International Water Management Institute.

Lindsay, S.W., Champbell, H., Adiamah, J.H., Greenwood, A.M., Bagali, J.E. & Greenwood, B.M. 1990, Malaria in a peri-urban area of the Gambia. Annals of Tropical and Medical Parasitology 84, 553-562.

Llamas, M. R. & Custodio, E. 2003, Intensive use of groundwater: A new situation which demands proactive action. In: *Intensive Use of Groundwater: Challenges and Opportunities*, edited by Llamas, R. and E. Custodio. A.A. Balkema, Rotterdam, Netherlands. pp.13-31.

Llamas, R., Back, W. & Margat, J. 1992, Groundwater use: Equilibrium between social benefits and potential environmental costs. Applied Hydrology 1 (2), 3-14, February.

Loeve, R., Molden, D., Dong, B., Li, Y.H., Chen, C.D., & Wang, J. Z. 2004, Issues of scale in water productivity in the Zhang He irrigation system: Implications for irrigation in the basin context. In: *Paddy and Water Environment* 2, edited by Randolph Barker (Guest Editor), 4, Dec. 2004.

Londhe, A., Talati, J., Singh, L. K, Vilayasseril, M., Dhaunta, S., Rawlley, B., Ganapathy, K. K. & Mathew, R. P. 2004, Urban-hinterland water transactions: a scoping study of six class I Indian cities. Paper presented at the IWMI Tata Annual Partners' Meet, Anand, 17-19 February 2004.

Maglinao, A.R. 1998, IBSRAM's impact: Making the difference in sustainable land management research. Issues in Sustainable Land Management 4. Bangkok, Thailand, International Board for Soil Research and Management. 41 pp.

Maglinao, A.R. & Leslie, R.N. (eds.). 2001, Soil Erosion Manageent Research in Asia Catchments: Methodological Approaches and Inital Results. *Proceedings of the 5th Management of Soil Erosion Consortium (MSEC) Assembly.* Thailand: IWMI Southeast Asa Regional Office. 275 pp.

Maglinao, A. R., Cagmat, R. B. & Sukristiyonubowo. 2003, Impact assessment: Methodology and its application to the Management of Soil Erosion Consortium (MSEC) Project. In: From soil research to land and water management: Harmonizing people and nature. Proceedings of the IWMI-ADB Project Annual Meeting and 7th Management of Soil Erosion Consortium (MSEC) Assembly, edited by A. R. Maglinao, C. Valentin and F. Penning de Vries. Thailand: IWMI Southeast Asia Regional Office. pp.209-220

Maglinao, A. R., Manzanilla, D. O. & Chandrapatya, S. 2005, Performance Evaluation and Impact Assessment of the ASIALAND Network: Management of Sloping Lands for Sustainable Agriculture Project-Phase 5, CD-ROM, IWMI-SEA.

Mainuddin, M. 2002, Groundwater irrigation in Bangladesh: Tool for poverty alleviation or cause of mass poisoning? Proceedings of the Symposium on Intensive Use of Groundwater: Challenges and Opportunities, Valencia, Spain, 10–14 December 2002.

MARA/ARMA. 1998, Towards and atlas of malaria risk in Africa. First technical report of the MARA/ARMA collaboration. MARA/ARMA initiative, Durban, South Africa (available at www.arma.org.za).

Masiyandima, M., Stoep, I. V. der, Mwanasawani, T. & Pfupajena, S. C. 2002, Groundwater management strategies and their implications on irrigated agriculture: The case of Dendron aquifer in Northern Province, South Africa. Physics and Chemistry of the Earth, Parts A/B/C 27 (11-22), 935-940.

Mati, B. Forthcoming, An Overview of Adaptable Technologies and Approaches in Rain-fed Agriculture in Eastern Africa. IWMI Working Paper. 100pp.

Matsuno, Y., Sakthivadivel, R. & Molden, D. 2000, Water saving irrigation in paddy rice: From field to basin levels. In: ICID, Asian Regional Workshop on Sustainable Development of Irrigation and Drainage for Rice Paddy Fields - *Proceedings, July 24th to 28th, 2000, Tokyo Japan.* Tokyo, Japan: ICID. Japanese National Committee. pp.227-236.

Matsuno, Y., Ko, H. S., Tan, C. H., Barker, R. & Levine, G. 2002, Accounting of Agricultural and Nonagricultural Impacts of Irrigation and Drainage Systems: A Study of Multifunctionality of Rice. IWMI Working Paper 43. Colombo, Sri Lanka, International Water Management Institute.

McKinney, D. C., Cai, X., Rosegrant, M. W., Ringler, C. & Scott, C. A. 1999, Modeling Water Resources Management at the Basin Level: Review and Future Directions. IWMI SWIM Paper 6. Colombo, Sri Lanka, International Water Management Institute.

Meher, R. 2003, Tail-enders and other deprived in canal irrigation systems: Orissa. Paper for the National Workshop on "Tail-enders and Other Deprived in Canal Irrigation Systems", organized by the Development Support Centre at Ahmedabad during November 28-29, 2003.

Meinzen-Dick, R. 1996, Groundwater Markets in Pakistan: Participation and Productivity. Washington, D.C., International Food Policy Research Institute.

Meinzen-Dick, R. & Mendonza, M. 1996, Alternative water allocation mechanisms: Indian and international experience. Economic and Political Weekly XXXI (13), March 30, pp: A25- A30.

Merrey, D.J. 1997, Expanding the Frontiers of Irrigation Management Research: Results of Research and Development at the International Irrigation Management Institute, 1984 to 1995. Colombo, Sri Lanka, International Water Management Institute, 228pp.

Merrey, D. J. 1996, Institutional Design Principles for Accountability in Large Irrigation Systems. IWMI Research Report 8. Colombo, Sri Lanka, International Water Management Institute.

Moench, M. 1992, Drawing down the buffer. Economic and Political Weekly 27(13), A7-A14.

Moench, M. 1994b, Approaches to groundwater management: To control or enable. Economic and Political Weekly September 24, A135-A146.

Moench, M. (ed.). 1994a, Selling water: Conceptual and policy debates over groundwater markets in India. Ahmedabad, India, VIKSAT.

Moench, M. & Burke, J. 2002, Groundwater and Society. Rome, Food and Agricultural Organization.

Moench, M., Burke, J. & Moench, Y. 2003, Rethinking the Approach to Groundwater and Food Security. Water Report 24. Rome, Food and Agriculture Organization.

Moench, M., Caspari, E. & Dixit, A. 1999, Rethinking the Mosaic: Investigations into Local Water Management. Kathmandu, Nepal: Nepal Water Conservation Foundation.

Molden, D. 1997, Accounting for Water Use and Productivity. IWMI SWIM Paper 1, Colombo, Sri Lanka, International Water Management Institute.

Molden, D. 2002, Meeting Water Needs for Food and Environmental Security. In: *Water for Sustainable Agriculture in Developing Regions – More Crop for Every Scarce. Proceedings of at the 8th JIRCAS International Symposium Series No.10, in Japan, 27-28 November 2001,* edited by Drop, M. Yajima, K. Okada and N. Matsumoto.

Molden, D. & de Fraiture, C. 2004, Blue Paper on Investing in Water for Food, Ecosystems and Livelihoods. Colombo, Sri Lanka, International Water Management Institute.

Molden D. J. & Sakthivadivel, R. 1999, Water Accounting to Assess Use and Productivity of Water. International Journal of Water Resources Development 15 (1-2): 55-71.

Molden, D., Amarasinghe, U. & Hussain, I 2001b, Water for Rural Development. Background Paper on Water for Rural Development Prepared for the World Bank. Working Paper 32. Colombo, Sri Lanka, International Water Management Institute.

Molden, D., Sakthivadivel, R. & Habib, Z. 2001a, Basin-Level Use and Productivity of Water: Examples from South Asia. IWMI Research Report 49. Colombo, Sri Lanka, International Water Management Institute.

Molden, J. D., Sakthivadivel, R. & Keller, J. 2001b, Hydronomic Zones for Developing Basin Water Conservation Strategies. IWMI Research Report 56. Colombo, Sri Lanka, International Water Management Institute.

Molden, D., Sakthivadivel, R., Samad, M. & Burton, M. 2005, Phases of river basin development: The need for adaptive institutions. In: *Irrigation and River Basin Management: Options for Governance and Institutions,* edited by Mark Svendsen. Wallingford, UK; Colombo, Sri Lanka: CABI; IWMI. pp.19-29.

Molden, D., Sakthivadivel, R., Perry, C. J., Fraiture, C. de & Kloezen, W. H. 1998, Indicators for Comparing Performance of Irrigated Agricultural Systems. IWMI Research Report 20. Colombo, Sri Lanka, International Water Management Institute.

Molle, F. 2003, Development Trajectories of River Basins: A Conceptual Framework. Research Report 72. Colombo, Sri Lanka, International Water Management Institute.

Molle, F., Shah, T. & Barker, R. 2003, The groundswell of pumps: Multilevel impacts of a silent revolution. Paper presented at the ICID-Asia Meeting, Taiwan, November 2003.

Montaigne, F. 2002, Water Pressure. National Geographic September 2002, 2-32.

Moriarty, P., Butterworth, J. & van Koppen, B. 2004, Beyond Domestic: Case Studies on Poverty and Productive Uses of Water at the Household Level. Technical Paper Series 41. International Water and Sanitation Centre, Delft, Netherlands. 241 pp.

Moya, P., Hong, L. Dawe, D. & Li, Y.H. 2001, Comparative assessment of on-farm water-saving irrigation techniques. In: Water Saving Irrigation for Rice, edited by R. Barker, R. Loeve, Y.H. Li and T.P. Tuong. Colombo, Sri Lanka, International Water Management Institute.

Mukherji, A. & Kishore, A. 2003, Tubewell Transfer in Gujarat: A Study of the GWRDC Approach. Research Report 69. Colombo, Sri Lanka, International Water Management Institute.

Mukherji, A. & Shah, T. 2002, Socio-ecology of groundwater irrigation in South Asia: An overview of issues and evidence. Proceedings of a Symposium on Intensively Exploited Aquifers (SINEX), Valencia, 10-14 December 2002.

Murray-Rust, D. H. & Svendsen, M. 2001, Performance of locally managed irrigation in Turkey: Gediz case study. Irrigation and Drainage Systems 15(4), 373-388.

Murray-Rust, H., Droogers, P. & Heydari, N. 2005, Water Allocation for the Future: Lining Irrigation and Water Allocation in the Zayandeh Rud Basin, Iran. Monograph. IWMI, Colombo, Sri Lanka, International Water Management Institute.

Murray-Rust, H., Lashari, B. & Memon, Y. 2000, Water Distribution Equity in Sindh Province, Pakistan. IWMI Working Paper 9. Colombo, Sri Lanka, International Water Management Institute.

Murray-Rust, H., Abdullaev, I., ul Hassan, M. & Horinkova, V. 2003, Water Productivity in the Syr-Darya River Basin. IWMI Research Report 67. Colombo, Sri Lanka, International Water Management Institute.

Murray-Rust, H., Sally, H. Salemi, H.R. & Mamanpoush, A. 2000, An Overview of the Hydrology of the Zayandeh Rud Basin. Iran-IWMI Research Report 3. Colombo, Sri Lanka, International Water Management Institute.

Nagar, R. K. 2002, The Groundwater Recharge Movement in Gujarat: A Quantitative idea of the Groundwater Recharge in Saurashtra and North Gujarat Regions: What Energized the Popular Movement?. Anand, India: IWMI-Tata Water Policy Program, Unpublished paper.

Namara, R. E., Upadhyay, B. & Nagar, R. K. 2005, Adoption and Impacts of Microirrigation Technologies: Empirical Results from Selected Localities of Maharashtra and Gujarat States of India. Research Report 93. Colombo, Sri Lanka, International Water Management Institute.

Namara, R. E., Weligamage, P. & Barker, R. 2003, Prospects for Adopting System of Rice Intensification in Sri Lanka: A Socioeconomic Assessment. Research Report No 75. Colombo, Sri Lanka, International Water Management Institute.

Neetha, N. 2004, Institutional choice in irrigation: A case study of distribution in a command area in Kerala. PhD thesis submitted to Jawaharlal Nehru University, New Delhi.

Nguyen-Khoa, S., Smith, L. & Lorenzen, K. 2005, Impacts of Irrigation on Inland Fisheries: Appraisals in Laos and Sri Lanka. Comprehensive Assessment Research Report 7. Colombo, Sri Lanka, Comprehensive Assessment Secretariat. vii, 36p.

Noble, A.D., Bossio, D.A., Penning de Vries, F.W.T., Pretty, J. & Thiyagarajan, T.M. In press. 'Bright Spots': Impacts and Drivers. Comprehensive Assessment of Water Management in Agriculture Research Report

Noble, A. D., Ruaysoongnern, S., Penning de Vries, F. W. T., Hartmann, C. & Webb, M. J. 2004, Enhancing the agronomic productivity of degraded soils in Northeast Thailand through clay-based interventions. In: Water in Agriculture: Proceedings of a CARDI International Conference "Research on Water in Agricultural Production in Asia for the 21st Century", Phnom Penh, Cambodia, 25-28 November 2003, edited by V. Seng, E. Craswell, S. Fukai. and K. Fischer. Canberra, Australia, ACIAR.

Noble, A.D., Berthelsen, S., Penning de Vries, F.W.T., Ruaysoongnern, S., Webb, M. & Giordano, M. 2004,. Enhancing the agronomic productivity of degraded soils through clay-based interventions. In prep.

Ohlsson, L. 1999, Environment, Scarcity and Conflict: A Study of Malthusian Concerns. Göteborg, Department of Peace and Development Research, University of Göteborg.

Ostrom, E. 1989, Governing the Commons: The Evolution of Institution for Collective Action. Workshop in Political Theory and Policy Analysis. Bloomington, IN, USA, 390pp.

Ostrom, E. 1992, Crafting Institutions for Self Governing Irrigation Systems. San Francisco, CA, USA, ICS Press. 111pp.

Ostrom, E. 1993, Design principles in long-enduring irrigation institutions. Water Resources Research 29(7), 1907-12.

Oweis, T. Y. & Hachum, A. Y. 2003, Improving water productivity in the dry areas of West Asia and North Africa. In: Water Productivity in Agriculture: Limits and Opportunities for Improvement, edited by J.W. Kijne, R. Barker and D. Molden. Wallingford, UK; Colombo, Sri Lanka: CABI; IWMI.

Oweis, T., Hachum, A. & Kijne, J. 1999, Water Harvesting and Supplemental Irrigation for Improved Water Use Efficiency in Dry Areas. IWMI SWIM Paper 7. Colombo, Sri Lanka, International Water Management Institute.

Palmer-Jones, R. 1995, Deep tube-well technology in South Asia. Norwich: Center for Development Studies, Draft for the Ford Foundation. (Unpublished)

Palmer-Jones, R.W. 1994, Groundwater markets in South Asia: A discussion of theory and evidence. In: Selling Water: Conceptual and Policy Debates over Groundwater Markets in India, edited by Marcus Moench, VIKSAT-Pacific Institute-Natural Heritage Institute.

Palmer-Jones, R.W. & Mandal, M.A.S. 1987, Irrigation Groups in Bangladesh. ODI/IIMI Irrigation Management Network, 87/2c, August. London, Overseas Development Institute.

Palmieri, A., Shah, F. & Dinar, A. 2001, Economics of reservoir sedimentation and sustainable management of dams. Journal of Environmental Management 61(2), 149-163.

Pant, N. 2004, Control of and Access to Groundwater in Uttar Pradesh. Paper presented in IWMI Tata Annual Partners' Meet, Anand, 17-19 February 2004.

Pardey, P.G. & J.E. Christian. 2002, The Production and Diffusion of Policy Knowledge: A Bibliometric Evaluation of the International Food Policy Research Institute. Impact Assessment Discussion Paper No. 14. Washington, DC: IFPRI.

Patil, S. K. & Doraiswamy, R. 2003, Tail-enders and other deprived in canal irrigation systems: Karnataka. Paper for The National Workshop on "Tail-enders and Other

Deprived in Canal Irrigation Systems", organized by Development Support Centre at Ahmedabad during November 28-29, 2003.

Penning de Vries, F.W.T. 2002, From soil research to land management: evolutions in the project Asialand sloping lands. In: *Management of Sloping Lands for Sustainable Agriculture,* edited by F.W.T. Penning de Vries. Final Report. Asialand Sloping Lands. Colombo, Sri Lanka, International Water Management Institute. 344 pp.

Penning de Vries, F.W.T., Siegert, K., Jesep, D. & Molanejad, M. (eds.). 2002a, Smallholder Land and Water Management Systems: Sharing Experiences in Africa, Asia and the Pacific. Paper presented at a seminar organized by IWMI, FAO, UN-ESCAP, Kasetsart University and Colombo Plan, Bangkok July 2002. Bangkok, Thailand, CD-ROM, IWMI.

Penning de Vries, F.W.T., Acquai, H., Molden, D., Scherr, S., Valentin, C. & Cofie, O. 2002b, Integrated Land and Water Management for Food and Environmental Security. Comprehensive Assessment Research Report 1. Colombo, Sri Lanka, International Water Management Institute, and Washington, D.C., Global Environmental Facility. 70 pp.

Perry, C. J. 1995, Alternative Approaches to Cost Sharing for Water Service to Agriculture in Egypt. Research Report No. 2. Colombo. International Irrigation Management Institute

Perry, C.J. 1996, The IIMI Water Balance Framework: A Model for Project Level Analysis. IWMI Research Report 5. Colombo, Sri Lanka, International Water Management Institute.

Perry, C.J. 2002, Charging for Irrigation Water: The Issues and Options, with a Case Study from Iran. Research Report No. 52. Colombo, Sri Lanka, International Water Management Institute.

Perry, C. J. & Narayanamurthy, S. G. 1998, Farmer Response to Rationed and Uncertain Irrigation Supplies. IWMI Research Report 24. Colombo, Sri Lanka, International Water Management Institute.

Perry, C. J., Rock, M. & Seckler, D. 1997, Water as an Economic Good: A Solution or a Problem? IWMI Research Report 14. Colombo, Sri Lanka, International Water Management Institute.

Phansalkar, S. & Verma, S. 2004, Improved water control as a strategy for enhancing tribal livelihoods - Outline of a proposal for a central India initiative. Paper presented in IWMI Tata Annual Partners' Meet, Anand, 17-19 February 2004.

Phommasack, T., Agus, F., Boosaner, A., Bricquet, J.-P., Chantavongsa, A., Chaplot, V., Ilao, R.O., Janeau, J.L., Marchand, P., Toan, T.D. & Valentin, C. 2002, Factorial Analysis of Runoff and Sediment Yield from Five Catchments and 21 Sub-Catchments of South-East Asia. In: MSEC Annual Report, edited by A. Maglinao. Colombo, Sri Lanka, International Water Management Institute.

Pinkowitz, L. 2002, Research dissemination and impact: Evidence from website downloads. Journal of Finance 57 (1): 485-499.

Pinstrup-Andersen, P. & Pandya-Lorch, R. 2001, The World Food Situation and 2020 Vision. Washington D.C., US, Food Policy Research Institute (IFPRI).

Postel, S. 1999, The Pillar of Sand: Can the Irrigation Miracle Last? Washington, The World Watch Institute.

Prathapar, S.A. & Qureshi, A.S. 1999, Mechanically Reclaiming Abandoned Saline Soils: A Numerical Evaluation. IWMI Research Report 30. Colombo. Sri Lanka, International Water Management Institute. 28 pp.

Pretty, J. & Hine, R. 2004, Measuring progress towards agricultural sustainability in developing regions. Report to DFID for project CNTR 02 4047: An Analysis of Findings from the University of Essex SAFE 2 Project.

Pretty, J., Noble, A., Bossio, D., Dixon, J., Hine, R., Morison, J. & Penning de Vries, F. 2004, In review. Resource-conserving agriculture increases yields in developing countries.

Puskur, Ranjitha & Thorpe, W. 2005, Crop and non-crop productivity gains: Livestock in water scarce watersheds. In: Watershed Management Challenges: Improving Productivity, Resources and Livelihoods, edited by Bharat Sharma, J.S. Samra, Christopher Scott and S.P. Wani. Colombo, Sri Lanka, International Water Management Institute. pp.95-115.

Qureshi, A. S. & Masih, I. 2002a, Modelling the effect of conjunctive water use on crops and environment. Paper presented in International Workshop on Conjunctive Water Management for Sustainable Irrigated Agriculture in South Asia, Lahore, Pakistan. 16-17 April 2002.

Qureshi, A. S. & Masih, I. 2002b, Managing soil salinity through conjunctive use of surface water and groundwater: A simulation study. Paper submitted for ICID Asian Regional Workshop on Sustainable Development of Water Resources and Operation and Management of Participatory Irrigation Organizations, Taipei Taiwan, 10-12 November 2003.

Qureshi, A. S., Murray-Rust H. & Masih, I. 2002b, An analytical approach to estimate groundwater pumpage. Proceedings, Second South Asia Water Forum, Islamabad Pakistan, 14-16 December 2002.

Qureshi, A. S., Turral, H. & Masih, I. 2004, Strategies for the Management of Conjunctive Use of Surface Water and Groundwater Resources in Semi-Arid Areas: A Case Study from Pakistan. Research Report No: 86. Colombo, Sri Lanka, International Water Management Institute. vi+24p.

Qureshi, A. S., Akhtar, M., Masih, I. & Bilal, M. 2002a, Sustaining groundwater boom: Protecting food security and livelihoods of rural poors in Pakistani Punjab. Paper presented at the Second South Asia Water Forum (SAWAF), Islamabad, 14-16 December 2002.

Rajagopal, A. 2003, Tail-enders and other deprived in canal irrigation systems: Tamilnadu, Paper for The National Workshop on "Tail-enders and Other Deprived in Canal Irrigation Systems", organized by Development Support Centre at Ahmedabad during November 28-29, 2003.

Raschid-Sally, L. 2000, Arsenic Contamination of Groundwater in Bangladesh. Discussion Paper iii. Colombo, Sri Lanka, International Water Management Institute.

Raskin, P., Gleick, P., Kirshen, P. Pontius, G. & Strzepek, K. 1997, Water Futures: Assessment of Long-Range Patterns and Prospects. Stockholm, Stockholm Environment Institute.

Rehman, G. 1997, GIS framework for salinity management across large irrigation systems: The Rechna interfluve in North Central Pakistan. ITIS 4(1), 21-23.

Rehman, G., Jehangir, W. A., Rehman, A., Aslam, M. & Skogerboe, G. V. 1997, Salinity management alternatives for the Rechna Doab, Punjab, Pakistan: Volume one - Principal findings and implications for sustainable irrigated agriculture. IIMI Pakistan report No. R-21.1. Lahore, Pakistan: IIMI Pakistan National Program. xii, 89p.

Renault, D. & Godaliyadda, G.G.A. 1999, Generic Typology for Irrigation Systems Operation. Research Report 29. Colombo, Sri Lanka, International Water Management Institute.

Renault, D. & Makin, I.W. 1999, Modernizing Irrigation Operations: Spatially Differentiated Resource Allocations. IWMI Research Report 35. Colombo, Sri Lanka, International Water Management Institute.

Renault, D. & Wallender, W. W. 2000, Nutritional water productivity and diets. Agricultural Water Management 45(3):275-296.

Rhoades, R.E. (ed.). 2001, Bridging Human and Ecological Landscapes; Participatory Research and Sustainable Development in an Andean Agricultural Frontier. Iowa, Kendal/Hunt Publishing Company. 368 pp.

Richard N., Boisvert, R & Hung H. C. Forthcomming, Multifunctional Agricultural Policy, Reduced Domestic Support, and Liberalized Trade: An Empirical Assessment for Taiwanese Rice. Colombo, Sri Lanka. Comprehensive Assessment Secretariat.

Rijsberman, F. 2003, Can development of water resources alleviate poverty? Water Policy 5(5-6).

Rijsberman, F.R. 2001, Can the CGIAR solve the world water crisis? Paper presented at the CGIAR Mid-Term meeting in Durban, South Africa, May 24.

Rijsberman, F.R. 2002, Troubled Waters, Water Troubles: Overcoming an Important Constraint to Food Security. In: In IFPRI, Sustainable food security for all by 2020: Proceedings of an international conference, Bonn, Germany, September 4-6, 2001. Washington, DC, USA. pp.140-145.

Rijsberman, F.R. 2004, Sanitation and access to clean water. In: Global Crises, Global Solutions, edited by D. Lomborg. Cambridge, UK, Cambridge University Press, pp. 498–540.

Rijsberman, F.R. 2006, Water scarcity: Fact or fiction? Agricultural Water Management 80 (2006), 5–22.

Rijsberman, F. R. (ed.). 2000, World Water Scenarios. London, Earthscan Publications.

Rijsberman, F. & de Silva, S. 2004, Water for food, health and livelihoods: Providing for the disadvantaged in times of scarcity. Paper presented at the 5th Annual Business Convention of the Ceylon Chamber of Commerce, at the Colombo Plaza, Colombo, Sri Lanka, 15 October 2004.

Rijsberman, F. & Mohammed, A. 2003, Water, food and environment: Conflict or dialogue? Water Science and Technology 47(6), 53-62.

Rijsberman, F.R. & Molden, D.J. 2001, Balancing Water Uses: Water for Food and Water for Nature. Thematic Background Papers. Bonn, International Conference on Freshwater, 3-7 December. Pp. 43-56.

Rockström, J. & Falkenmark, M. 2000, Semiarid crop production from a hydrological perspective – gap between potential and actual yields. Critical Reviews of Plant Science 19 (4), 319-346.

Rockström, J., Barron, J., & Fox, P. 2003, Water productivity in rain-fed agriculture: Challenges and opportunities for smallholder farmers in drought-prone tropical agroecosystems. In: Water Productivity in Agriculture: Limits and Opportunities for Improvement, edited by. J.W. Kijne, R. Barker and D. Molden. Wallingford (UK), CABI Publishing, pp. 145-162.

Rockstrom, J., Gordon, L., Falkenmark, M., Folke, C. & Engvall, M. 1999, Linkages among water vapor flows, food production, and terrestrial ecosystem services. Conservation Ecology 3(2), 5 (http://www.

consecol. org/vol3/iss2/art5).

Rogers, P. & Hall, A.W. 2003, Effective Water Governance. TEC Background Papers No.7. Stockholm, Global Water Partnership.

Rogers, P., Bhatia, R. & Huber, A. 1997, Water as a Social and Economic Good: How to Put the Principle into Practice. TAC Background Papers No. 2. Available on line at (http://www.gwpforum.org/gwp/library/ Tac2.pdf/)

Rosegrant, M.W. & Gazmuri, R. S. 1994, Reforming Water Allocation Policy through Markets in Tradable Water Rights: Lessons from Chile, Mexico and California. EPTD Discussion Paper No.6. Washington, D.C., International Food Policy Research Institute.

Rosegrant, M.W., Cai, X. & Cline, S. 2002, World Water and Food to 2025: Dealing with Scarcity. Washington, D.C., International Food Policy Research Institute.

Roy, A. D. & Shah, T. 2002, The Socio-ecology of Groundwater in India. Water Policy Briefing No. 4. Colombo, Sri Lanka, International Water Management Institute.

Roy, A.D. & Shah, T. 2003, Socio-ecology of Groundwater Irrigation in India. In: Intensive Use of Groundwater: Challenges and Opportunities, edited by R. Llamas and E. Custodio. The Netherlands, Swets and Zetlinger Publishing Co.

Roy, K. C. & Mainuddin, M. 2003, Socio-ecology of groundwater irrigation in Bangladesh. Paper presented in workshop on Groundwater Socio-Ecology of Asia: Governing a colossal anarchy. Anand, India, IWMI-Tata Water Policy Program, Anand, India, 28 January 2003.

Saeed, M. M., Asghar, M. N. & Bruen, M. 2003, Options for skimming fresh groundwater in the Indus Basin of Pakistan: A review. Journal of Groundwater Hydrology 45 (3).

Sakthivadivel, R. & Gulati, O.T. 1997, Assessing Management Interventions Impact with IIMI's Minimum Set of Comparative Indicators: The Case of Mahi-Kadana Project, India. In IIMI; ILRI; INA - CRA; IHE; UNC, International seminar. Research Program on Irrigation Performance (RPIP), Mendoza, Argentina, November 3-7, 1997. 14p.

Sakthivadivel, R. & Molden, D. 2002, Linking water accounting analysis to institutions: Synthesis of five country studies. In: Integrated Water-Resources Management in a River Basin Context: Institutional Strategies for Improving Productivity of Agricultural Water Management. Proceedings of the Regional Workshop, Malang, Indonesia, January 15-19, 2001, edited by B. Bruns, D. J. Bandaragoda and M. Samad. Colombo, Sri Lanka, International Water Management Institute. 315pp.

Sakthivadivel, R. & Nagar, R. K. 2003, Private Initiative for Groundwater Recharge: Case of Dudhada Village in Saurashtra. Anand, IWMI-Tata Research Highlight No. 15.

Saktivadivel, R. & Sally, H. 2003, Irrigation Management for the Tropics. Encyclopedia of Water Science. Dekker.

Sakthivadivel, R., Fernando, N. & Brewer, J.D. 1997, Rehabilitation Planning for Small Tanks in Cascades: A Methodology Based on Rapid Assessment. IIMI Research Report 13. Colombo, Sri Lanka, International Water Management Institute.

Sakthivadivel, R., Thiruvengadachari, S. & Amarasinghe, U.A. 1998, Intervention analysis of an irrigation system using a structured system concept. In: Modernization of irrigation system operations. ITIS 5 - Information Techniques for Irrigation Systems: Proceedings of the Fifth International ITIS Network Meeting,

Aurangabad, Maharashtra, India, 28-30 October 1998, edited by D. Renault, pp.43-65.

Sakthivadivel, R., Thiruvengadachari, S. & Amarasinghe, U. A. 1999a, Modernization Using the Structured System Design of the Bhadra Reservoir Project, India: An Intervention Analysis. IWMI Research Report 33. Colombo, Sri Lanka, International Water Management Institute.

Sakthivadivel, R., Amarasinghe, U. A. & Thiruvengadachari, S. 2001, Using Remote Sensing Techniques to Evaluate Lining Efficacy of Watercourses. IWMI Research Report 46. Colombo, Sri Lanka, International Water Management Institute.

Sakthivadivel, R., Loeve, R., Amarasinghe, U. A. & Hemakumara, M. 2001, Water Scarcity and Managing Seasonal Water Crisis: Lessons from the Kirindi Oya Project in Sri Lanka. IWMI Research Report 55. Colombo, Sri Lanka, International Water Management Institute.

Sakthivadivel, R., Brewer, J.D., Bhatia, R., Narayanamurthy, S. & Rao, P.S. 1995a, Evaluation of Schemes under NWMP-1: Bhadra Scheme in Karnataka and Sathanur Scheme in Tamil Nadu. Volume I - Main Report. Final Evaluation Report under India's National Water Management Project I. ix, 53p. + annex.

Sakthivadivel, R., Brewer, J.D., Bhatia, R., Narayanamurthy, S. & Rao, P.S. 1995b, Evaluation of Schemes under NWMP-1: Bhadra Scheme in Karnataka and Sathanur Scheme in Tamil Nadu. Volume II - Scheme Report. Final Evaluation Report under India's National Water Management Project I. xvi, 177p. + annexes.

Sakthivadivel, R., Fraiture, C. de, Molden, D.J., Perry, C. & Kloezen, W. 1999b, Indicators of land and water productivity in irrigated agriculture. International Journal of Water Resources Development 15(1/2), 61-179.

Sakthivadivel, R., Thiruvengadachari, S., Amerasinghe, U., Bastiaanssen, W.G.M. & Molden, D. 1999, Performance Evaluation of the Bhakra Irrigation System, India, Using Remote Sensing and GIS Techniques. IIMI Research Report 28. Colombo, Sri Lanka, International Water Management Institute.

Saleth, R. M. 1994, Groundwater markets in India: A legal and institutional perspective. The Indian Economic Review 39(2), 157-176.

Saleth, R. M. 2004, Strategic Analysis of Water Institutions in India: Application of a New Research Paradigm. Research Report 79. Colombo Sri Lanka, International Water Management Institute.

Saleth, R. M. & Dinar, A. 2004, The Institutional Economics of Water: A Cross-country Analysis of Institutions and Performance. Cheltenham, UK, Edward Elgar Publishing Ltd.

Saleth, R. M. & Dinar, A. 2005, Water institutional reforms: Theory and practice. Water Policy (Special Issue) 7, 1-19.

Saleth, R. M., Namara, R. E. & Samad, M. 2003, Dynamics of irrigation-poverty linkages in Rural India: Analytical framework and empirical analysis. Water Policy 5(5/6), 459-473.

Saleth, R. M., Samad, M., Molden, D. & Hussain, I. 2003, Water, poverty, gender: An overview of issues and policy implications, Water Policy (Special Issue) 5(5/6), 385-398.

Sally, H., Saktivadivel, R. & Molden, D.J. 2000, More crop per drop: Considerations for precision irrigation in a basin context. In: Micro-irrigation Technologies for Developing Agriculture. Congress S. Africa. October 2000.

Samad, M., Dingle, M.A. & Shafique, M.S. 1995, Political and economic dimensions of privatization and turnover of irrigation schemes in Sudan. In: *Irrigation*

Management Transfer: Selected Papers from the International Conference on Irrigation Management Transfer, Wuhan, China, 20-24 September 1994, edited by S. H. Johnson, D. L. Vermillion and J. A. Sagardoy. Rome, Italy: FAO, pp.211-223; pp.457-476.

Sarwar, A. & Bastiaanssen, W. G. M. 2001, Long-term effects of irrigation water conservation on crop production and environment in semiarid areas. Journal of Irrigation and Drainage Engineering 127(6), 331-338.

Sayer, J. & Campbell, B. 2002. Research to integrate productivity enhancement, environmental protection and human development. Conservation Ecology, vol 5 (2).

Scherr, S.J. & Yadav, S. 1996, Land Degradation in the Developing World: Implications for Food, Agriculture, and the Environment to 2020. Food, Agriculture, and Environment Discussion Paper 14. Washington, D.C. (USA), International Food Policy Research Institute.

Scott, C. 2003, Sustainable groundwater management: Have property rights reforms helped in Mexico? In: *Groundwater Governance in Asia: The Challenge of Taming a Colossal Anarchy*, edited by IWMI. International Water Management Institute.

Scott, C. A., Faruqui, N.I. & Raschid-Sally, L. 2004, Wastewater Use in Irrigated Agriculture: Confronting the Livelihood and Environmental Realities. Wallingford: CABI/International Water Management Institute/International Development Research Centre.

Scott, A. C., Zarazua J. A. & Levine, G. 2000, Urban-Wastewater Reuse for Crop Production in the Water-Short Guanajuato River Basin, Mexico. IWMI Research Report 41. Colombo, Sri Lanka, International Water Management Institute.

Seago, C. 2003, A Guide to Manufacturing the Kit Treadle Pump. Private bag X519, Silverton 0127, S. Africa. The ARC-Institute for Agricultural Engineering, 22 pp.

Seckler, D. 1996, The New Era of Water Resources Management: From "Dry" to "Wet" Water Savings. IWMI Research Report 1. Colombo, Sri Lanka, International Water Management Institute.

Seckler, D., Amarasinghe, U., Molden, D., de Silva, R. & Barker, R. 1998a, World Water Demand and Supply, 1990 to 2025: Scenarios and Issues. IIMI Research Report 19. Colombo, Sri Lanka, International Irrigation Management Institute. vi, 40p.

Seckler, D., Molden, D. & Barker, R. 1998b, Water scarcity in the twenty-first century. In: IWMI Sri Lanka, 6th MANCO (Mancid Annual Conference) - "Sustainable Rice Production". World Water Demand and Supply 1990-2025: Scenarios and Issues, 13 October 1998. IWMI Water Brief no.1. Alor Setar, Kedah. 12p.

Seckler, D., Molden, D. & Sakthivadivel, R. 2003, The concept of efficiency in water-resources management and policy. In: *Water Productivity in Agriculture: Limits and Opportunities for Improvement,* edited by J.W. Kijne, R. Barker & D. Molden. 2003. Wallingford, UK, CABI Publishing, pp. 37-51.

Seckler, D., Sakthivadivel, R., Keller, A., Amarasinghe, U.A., Fraiture, C. de & Molden, D. 2000, World Water Supply and Demand: 1995 to 2025. Colombo, Sri Lanka: International Water Management Institute.

Sengtaheunghoung, O. 2002, Soil organic matter and soil fertility in the long term experiments in Laos PDR. In: *Management of Sloping Lands for Sustainable Agriculture: ASIALAND Sloping Land Phase 4*, edited by F. Penning de Vries. Colombo, Sri Lanka, International Water Management Institute. pp. 95-109.

Shah, A. 2003, Tail-enders and Other Deprived in Canal Irrigation Systems: Gujarat, Paper for The National Workshop on "Tail-enders and Other Deprived in Canal

Irrigation Systems", organized by Development Support Centre at Ahmedabad during November 28-29, 2003.

Shah, T. 1993, Groundwater Markets and Irrigation Development: Political Economy and Practical Policy. Bombay, India, Oxford University Press.

Shah, T. 2000, Mobilizing social energy against environmental challenge: Understanding the groundwater recharge movement in western India. Natural Resource Forum 24 (2000), 197-209.

Shah, T. 2003a, Governing the Groundwater Economy: Comparative Analysis of National Institutions and Policies in South Asia, China and Mexico. Water Perspectives 1(1).

Shah, T. 2003b, Wells and welfare in Ganga basin: Public policy and private initiative in eastern Uttar Pradesh, India. IWMI Research Report No. 54; also published in *Water Resources and Sustainable Development*, edited by Kamta Prasad. New Delhi, Shipra Publications. IRMED

Shah, T. 2004, Community management (Comman) of groundwater resources: An appropriate response to groundwater overdraft in India? Comment on Comman Project Report.

Shah, T. & Desai, R. 2002, Creative destruction: Is that how Gujarat is adapting to groundwater depletion? A synthesis of ITP studies. Anand, India: IWMI-Tata Water Policy Research Program.

Shah, T. & Keller, J. 2002, Micro-irrigation and the poor: A marketing challenge in small holder irrigation development. In: *Private Irrigation in sub-Saharan Africa: Proceedings of a Regional Seminar on Private Sector Participation in Irrigation in sub-Saharan Africa,* edited by Hilmy Sally and Charles Abernethy. Colombo, Sri Lanka: International Water Management Institute. pp. 165-183.

Shah, T. & Singh, O. P. 2003, Irrigation and tribal poverty in India: Improved water control as the strategy for agricultural intensification in India's Tribal Heartland. Paper presented at IWMI Tata Annual Partners' Meet, Anand, 26-29 January 2003.

Shah, T. & Singh, O.P. 2003, Can Irrigation Eradicate Rural Poverty in Gujarat? Water Policy Research 10. Gujarat, India, IWMI-Tata Water Policy Program.

Shah, T. & Singh, O. P. 2004, Irrigation development and rural poverty in Gujarat, India. A disaggregated analysis. Water International 29(2), 167-177.

Shah, T., Giordano, M. & Wang, J. 2004, Irrigation institutions in a dynamic economy: What is China doing differently from India? Economic and Political Weekly 39(31), 3452-3461.

Shah, T., Hussain, I. & Rehman, S. U.R. 2000, Irrigation Management in Pakistan and India: Comparing Notes on Institutions and Policies. Working Paper No. 4. Colombo, Sri Lanka, International Water Management Institute.

Shah, T., Makin, I. & Sakthivadivel. R. 2001, Limits to leapfrogging: Issues in transposing successful river basin management institutions in the developing world. In: *Intersectoral Management of River Basins,* edited by Charles L. Abernethy. International Water Management Institute.

Shah, T., Scott, C. & Buechler, S. 2004, A decade of water sector reforms in Mexico: Lessons for India's new water policy. Economic and Political Weekly 39(4), 361-370.

Shah, T., Singh, O. P. & Mukherji, A. 2006, Some aspects of South Asia's groundwater irrigation economy: Analyses from a survey in India, Pakistan, Nepal terai and Bangladesh. Hydrogeology Journal. (In press)

Shah, T., Molden, D., Sakthivadivel, R. & Seckler, D. 2001, Global groundwater situation: Opportunities and challenges. Economic and Political Weekly, 36(43): 4142-4150.

Shah, T., Roy, A. D., Qureshi, A.S. & Wang, J. 2003, Sustaining Asia's groundwater boom: An overview of issues and evidence. Natural Resources Forum 27(2003), 130-140.

Shah, T., Scott, C., Kishore, A. & Sharma, A. 2004, Energy-Irrigation Nexus in South Asia: Improving Groundwater Conservation and Power Sector Viability. Research Report 70. Colombo, Sri Lanka, International Water Management Institute.

Shah, T., Alam, M., Kumar, M. D., Nagar, R.K. & Singh, M. 2001, Pedaling out of Poverty: Social Impact of a Manual Irrigation Technology in S. Asia. IWMI Research Report 45. Colombo, Sri Lanka, International Water Management Institute. 37 pp.

Shah, T., Koppen, B. van, Merrey, D., Lange, M. de & Samad, M. 2002, Institutional Alternatives in African Smallholder Irrigation: Lessons from International Experience with Irrigation Management Transfer. IWMI Research Report 60. Colombo, Sri Lanka, International Water Management Institute.

Shankar, K. 1992, Dynamics of Groundwater Irrigation. New Delhi, Segment Books.

Shiklomanov, I. 2000, World water resources and water use: Present assessment and outlook for 2025. In: World Water Scenarios: Analyses, edited by Frank R. Rijsberman. London, UK, Earthscan Publications Ltd.

Shingi, P. & Asopa, V. N. 2002, Independent Evaluation of Checkdams in Gujarat: Strategies and Impacts. Ahmedabad, India, Indian Institute of Management.

SIMI (Smallholder Irrigation Market Initiative). 2003, Smallholder Irrigation Market Initiative. http://www.siminet.org/fs_start.htm

Smakhtin, V. U. 2003, Environmental water needs and impacts of irrigated agriculture in river basins: A framework for new research program. In: Training Center of the Intergovernmental Water Committee of the Middle Asia. Ecological Water Releases, Issue 1. pp.31-41.

Smakhtin, V., Revenga, C. & Döll, P. 2004, Taking into Account Environmental Water Requirements in Global-Scale Water Resources Assessments. Comprehensive Assessment Research Report 2, Colombo, Sri Lanka, Comprehensive Assessment Secretariat.

Smaling, E.M. 1993, An agro-ecological framework for integrated nutrient management. PhD thesis. Wageningen, Netherlands, Wageningen Agricultural University. 250 pp.

Smyth, A.J. & Dumanski, J. 1993, FESLM: An International Framework for Evaluating Sustainable Land Management. FAO World Soil Resources Report 73. Rome, FAO.

Stanzel, P., Oze, A., Smakhtin, V., Boelee, E. & Droogers, P. 2002, Simulating Impacts of Irrigation on the Hydrology of the Karagan Lagoon in Sri Lanka. IWMI Working Paper 44. Colombo, Sri Lanka, International Water Management Institute.

Steenbergen, F. V. & Shah, T. 2003, Rules rather than rights: Self-regulation in intensively used groundwater systems. In: Intensive Use of Groundwater: Challenges and Opportunities, edited by R. Llamas and E. Custodio. The Netherlands, Swets and Zetlinger Publishing Co.

Stimie, C., Richters, E., Thompson, H., Perret, S., Matete, M., Abdallah, K., Kau, J. & Mulibana, E. 2001, Hydro-Institutional Mapping in the Steelpoort River Basin, South Africa. IWMI Working Paper 17. Colombo, Sri Lanka, International Water Management Institute.

Strosser, P. & Kuper, M. 1994, Water Markets in the Fordwah/Eastern Sadiquia Area: An Answer to Perceived Deficiencies in Canal Water Supplies? Working Paper No. 10. Colombo, Sri Lanka, International Irrigation Management Institute.

Strosser, P. & Meinzen-Dick, R. 1994, Groundwater markets in Pakistan: An analysis of selected issues. In: Selling Water: Conceptual and Policy Debates over Groundwater Markets in India, edited by Marcus Moench. VIKSAT-Pacific Institute-Natural Heritage Institute.

Suzuki, S., Noble, A. D., Ruaysoongnern, S. & Chinabut. N. In prep. Improvement in water holding capacity and structural stability of a degraded light textured sandy soil in Northeast Thailand.

Svendsen, M. (ed.). 2005, Irrigation and River Basin Management: Options for Governance and Institutions. Oxford, UK, CAB International and IWMI. pp 258.

Svendsen, M. & Murray-Rust, D. H. 2001, Creating and consolidating locally managed irrigation in Turkey: The national perspective. Irrigation and Drainage Systems 15(4), 355-371.

Syers, K. & Bouma, J. (eds.). 1998, Resource management domains. IBSRAM Proceedings 16. Bangkok, Thailand, International Board for Soil Research and Management. 322 pp.

Tahir, Z. & Zaigham Habib. 2000, Land and Water Productivity: Trends across Punjab Canal Commands. IWMI Working Paper 14. Colombo, Sri Lanka, International Water Management Institute.

Tewari, D. D. 2003, Groundwater socio-ecology of the Limpopo province, South Africa: A preliminary study. Unpublished paper.

Thobani, M. 1995, Tradable Property Rights to Water: How to Improve Water Use and Resolve Water Conflicts. Viewpoint, Washington, The World Bank, FPD Note No 34, February.

Tren, R. & Schur, M. 2000, Olifants River Irrigation Schemes. Reports 1 & 2. IWMI Working Paper 3. Colombo, Sri Lanka, International Water Management Institute.

Tuinhof, A. & Heederik, J. P. 2002, Management of Aquifer Recharge and Subsurface Storage. Netherlands National Committee and International Association of Hydro-Geologists.

Turral, H. 1998, Hydrologic? Reform in Water Resources Management in Developed Countries with Major Agricultural Water Use. London: Overseas Development Institute.

Tyagi, N. K, Agrarwal, A, Sakthivadivel R. & Ambast, S.K. 2003, Managing Scarce Water Supply. An Analysis of the Farmers Decision. World Water Congress XII, International Water Resources Association.

UNWWDR (United Nations World Water Development Report). 2003, Water for People, Water for Life. Paris, UNESCO.

Upadhyay, B. 2003, Drip Irrigation: An Appropriate Technology for Women. Appropriate Technology. December, 2003.

Upadhyay, B. 2003, Water, poverty and gender: Review of evidences from Nepal, India and South Africa. Water Policy 5(5-6), 503-511.

Upadhyay, B., Samad, M. & Giordano, M. 2005, Livelihoods and Gender Roles in Drip-irrigation Technology: A Case of Nepal, Working paper No. 87. Colombo, Sri Lanka, International Water Management Institute.

Uphoff, N. 1986, Improving International Irrigation Management with Farmer Participation: Getting the Process Right. Boulder, CO, USA, Westview Press, 215pp.

Vallee, D. 2005, Dialogue on Water, Food and Environment Progress Report: November 2003-June 2005. Unpublished Report. July 2004. Colombo, Sri Lanka, Dialogue Secretariat.

Van der Hoek, W., Sakthivadivel, R., Renshaw, M., Silver, J.B., Birley, M.H. & Konradsen, F. 2001, Alternate Wet/Dry Irrigation in Rice Cultivation: A Practical Way to Save Water and Control Malaria and Japanese Encephalitis? IWMI Research Report 47. Colombo, Sri Lanka, International Water Management Institute.

Van Etten, J., Koppen, B. van., & Pun, S. 2002, Do Equal Land and Water Rights Benefit the Poor? Targeted Irrigation Development: The Case of the Andhi Khola Irrigation Scheme in Nepal. IWMI Working Paper 38. Colombo, Sri Lanka, International Water Management Institute.

van Koppen, B. 2000, From Bucket to Basin: Managing River Basins to Alleviate Water Deprivations. Colombo, Sri Lanka, International Water Management Institute.

van Koppen, B. 2002, A Gender Performance Indicator for Irrigation: Concepts, Tools and Applications. IWMI Research Report 59. Colombo, Sri Lanka, International Water Management Institute.

Vashishtha, P., Deva, V. & Kumar, Y. 2003, Tail-enders and other deprived in canal irrigation systems: Haryana. Paper for The National Workshop on "Tail-enders and Other Deprived in Canal Irrigation Systems", organized by Development Support Centre at Ahmedabad during November 28-29, 2003.

Verma, S. 2003, More crop per drop: Can micro-irrigation help alleviate groundwater depletion?, In: Groundwater Governance in Asia: The Challenge of Taming a Colossal Anarchy, edited by IWMI. International Water Management Institute.

Verma, S. 2004, Promoting micro irrigation in India: A review of evidence and recent developments. Paper presented at IWMI Tata Annual Partners' Meet, Anand, 17-19 February 2004.

Verma, S., Tsephal, S. & Jose, T. 2004, Pepsee systems: Grassroots innovation under groundwater stress. Water Policy 6(4), 303-318.

Vermillion, D.L. 1997, Impacts of Irrigation Management Transfer: A Review of the Evidence. Research Report 11. Colombo, Sri Lanka, International Water Management Institute.

Vermillion, D.L. 1998, Impacts of irrigation management transfer: Results from IIMI's research. INPIM 7(1-3), 12-13, 21.

Vermillion, D. L. & Garcés-Restrepo, C. 1996, Results of Management Transfer in Two Irrigation Districts in Colombia. IIMI Research Report 4. Colombo, Sri Lanka, International Irrigation Management Institute.

Vermillion, D. L., Samad, M., Amerasinghe, U. & Narayanamurthy, S.G. 1996, A Standard Methodology to Assess the Impacts of Irrigation Management Transfer. Colombo, Sri Lanka, International Irrigation Management Institute.

Vermillion, D., Samad, M., Pusposutardjo, S., Arif, S.S. & Rochdyanto, S. 2000, An Assessment of Small Scale Irrigation Management Transfer Program in Indonesia. IWMI Research Report 38. Colombo, Sri Lanka, International Water Management Institute.

Wade, R. 1982, The system of administrative and political corruption: Canal irrigation in south India. Journal of Development Studies 18 (3), 287-328.

Wallace, J. S. 2000, Increasing agricultural water use efficiency to meet future food production. Agriculture, Ecosystems and Environment 82 (2000) 105–119.

Wang, J. 2003, Sustainable groundwater management: How effective has groundwater regulation been in North China Plains? In: Groundwater Governance in Asia: The Challenge of Taming a Colossal Anarchy, edited by IWMI. International Water Management Institute.

Wang, J., Huang, J. & Rozelle, S, 2000, Property Right Innovation and Technical Efficiency – Case Study of Groundwater Irrigation in Hebei, China. Working Paper #wp-00-e22. Center for Chinese Agricultural Policy.

Wester, P., Merrey, D.J. & de Lange, M. 2003, Boundaries of consent: Stakeholder representation in river basin management in Mexico and South Africa. World Development 31(5), 797-812.

Wester, P., Pimentel, B. M. & Scott, C. 1999, Institutional Responses to Groundwater Depletion: The Aquifer Management Councils in the State of Guanajuato. Colombo, Sri Lanka: International Water Management Institute (internal draft). Presented at the International Symposium on Integrated Water Management in Agriculture, Gomez Palacio, Mexico, June 16-18.

WHO (World Health Organization). 2003, Water Sanitation and Health Website: (http://www.who.int/water_sanitation_health/hygiene/en/).

WHO (World Health Organization). 2006, Guidelines for the safe use of wastewater, excreta and grey water: Wastewater use in agriculture, Volume 2, Geneva *(http://www.who.int/water_sanitation_health/wastewater/gsuweg2/en/index.html).*

Wood, G. 1995, Private Provision after Public Neglect: Opting Out with Pumpsets in North Bihar. Centre for Development Studies. Bath, University of Bath.

Wood, S., Sebastian, K. & Scherr, S.J. 2000, Soil Resource Condition. In: Pilot Analysis of Global Ecosystems (PAGE). Washington, IFPRI and World Resources Institute. pp 45-54.

World Bank. 2003, The Water Resources Sector Strategy: An Overview. Washington, D.C., The World Bank.

World Bank. 2004, Drain for Gain: Integrated Solutions to Drainage in Land and Water Management. Agricultural, and Rural Development. Washington D.C., World Bank.

Xiang, Q., Huang, J. & Wang, J. 2000, Property Right Innovation and Cropping Pattern Adjustment: Case Study of Groundwater Irrigation System in Hebei, China. Center for Chinese Agricultural Policy (draft), Beijing.

Yu, B. & Rose, C.W. 1999, Application of a physically based soil erosion model in the absence of data on runoff rates: I. Theory and methodology. Aust. J. Soil Research 37 (1).

Yu, B., Sajjapongse, A., Dixin, Y., Eusof, Z., Aneksamphant, C., Rose, C. & Cakurs, U. 1999, Application of a physically based soil erosion model in the absence of data on runoff rates: II. Four case studies from China, Malaysia, and Thailand. Aust. J. Soil Research 37 (1).

Zwart & Bastiaanssen. 2004, Review of measured crop water productivity values for irrigated wheat, rice, cotton and maize. Agricultural Water Management 69, 115-133.

Zwarteveen, M.Z. 1997, A Plot of One's Own: Gender Relations and Irrigated Land Allocation Policies in Burkina Faso. IWMI Research Report 10, Colombo, Sri Lanka, International Water Management Institute.

Zwarteween, M. Z & Neupane, N. 1996, Free-riders or victims: Women's nonparticipation in irrigation management in Nepal's Chhattis Mauja irrigation scheme. Research Report 7. Colombo, Sri Lanka: International Irrigation Management Institute.

Annex A

List of Main Donors

(1) African Development Bank (AfDB)
(2) Asian Development Bank (ADB)
(3) Australian Centre for International Agricultural Research (ACIAR)
(4) Bundesministerium für Wirtschaftliche Zusammenarbeit und Entwicklung (BMZ), Germany
(5) Canadian International Development Agency (CIDA)
(6) CARE International
(7) CEMEGREF
(8) Danish International Development Agency (DANIDA), Denmark
(9) Department For International Development (DFID), UK
(10) Deutsche Gesellschaft für Technische Zusammenarbeit (GTZ), Germany
(11) European Union
(12) Global Environment Facility (GEF)
(13) Government of Belgium
(14) Government of Cambodia
(15) Government of China
(16) Government of France
(17) Government of India
(18) Government of Iran
(19) Government of Ireland
(20) Government of Israel
(21) Government of Nepal
(22) Government of Netherlands
(23) Government of Norway
(24) Government of Pakistan
(25) Government of South Africa
(26) Government of Sri Lanka
(27) Government of Taiwan
(28) Government of Thailand
(29) International Development Research Centre (IDRC), Canada
(30) International Fund for Agricultural Development (IFAD)
(31) Japan Bank for International Cooperation (JBIC)
(32) Japan International Cooperation Agency (JICA)
(33) National Oceanic & Atmospheric Authority (NOAA), USA
(34) New Partnership for Africa Development (NEPAD)
(35) OPEC Fund for International Development, Austria
(36) Sir Ratan Tata Trust, India
(37) Swedish International Development Cooperation Agency (SIDA)
(38) Swiss Agency for Development and Cooperation (SDC)
(39) United Nations Educational Scientific & Cultural Organization (UNESCO)
(40) United Nations Food and Agriculture Organization (FAO)

(41) US Agency for International Development (USAID), USA
(42) Water and Power Development Authority (WAPDA), Pakistan
(43) World Bank (WB)
(44) World Health Organization (WHO)

Annex B

List of Key Partners and Collaborators

Virtually, all IWMI's projects now involve a range of partners that are, more or less, involved in the research, outreach or capacity building. Many projects are undertaken through consortia or research networks that have large numbers of partners. Each IWMI office in each country or region has its own network of local partners. A program like the IWMI-Tata Water Policy program works with dozens of partners. The Comprehensive Assessment and the Challenge Program on Water and Food, two programs IWMI has initiated and hosts, each has literally hundreds of partners, but some of these are not partners that IWMI works with directly. It is, therefore, not possible to provide a complete list of all of IWMI's collaborators and partners over the period this book covers. This Annex gives a somewhat arbitrary list (in alphabetical order) of IWMI's key partners in 2005 (not including the other CGIAR centers or donor organizations), with apologies to the many other organizations we work with that are not listed here.

North Partners and Collaborators

(1) Center for Development Research (ZEF), University of Bonn, Germany
(2) Food and Agricultural Organization of the United Nations (FAO)
(3) Global Water Partnership (GWP), Sweden
(4) Institut de Recherche pour le Développement (IRD), France
(5) UNESCO-IHE International Institute for Water Education, the Netherlands
(6) International Programme for Technology and Research in Irrigation and Drainage (IPTRID), Italy
(7) International Reference Center for Water and Sanitation (IRC), Netherlands
(8) Ramsar Convention Bureau and Secretariat, Switzerland
(9) Stockholm Environment Institute (SEI), Sweden
(10) Stockholm International Water Institute (SIWI), Sweden
(11) Wageningen Agricultural University, The Netherlands
(12) Winrock International
(13) World Health Organization (WHO), Department of Water Supply and Sanitation, Switzerland

South Partners and Collaborators

(1) Agricultural Research & Education Organization (AREO), Iran
(2) Aga Khan Rural Support Programme I (AKRSP), India
(3) Anti Malaria Campaign, Ministry of Health, Sri Lanka
(4) Central Environmental Authority (CEA), Sri Lanka
(5) Centre for Soil and Agroclimate Research (CSAR), Indonesia
(6) Chinese Center for Agricultural Policy (CCAP), China
(7) Council for Scientific and Industrial Research (CSIR), Ghana

(8) Department of Irrigation Punjab, Pakistan
(9) Department of Irrigation, Nepal
(10) Department of Water Affairs and Forestry (DWAF), South Africa
(11) Dhan Foundation (DF), India
(12) Environment Protection Training and Research Institute (EPTRI), India
(13) Forest, Rangeland and Watershed Management Organization (FRWO), Iran
(14) Ghana Irrigation Development Authority (GIDA), Ghana
(15) Gulistan University of Uzbekistan
(16) Gidrogeo Institute of Uzbekistan
(17) Indian Council of Agriculture Research (ICAR), India
(18) Institute of Public Health (IPH), Pakistan
(19) Institute of Rural Management (IRM), India
(20) Institute of Water and Human Resource Development (IWHRD), Nepal
(21) Irrigation & CAD Department, Government of Andhra Pradesh, India
(22) Jawaharlal Nehru Technical University (JNTU), India
(23) Khon Kaen University (KKU), Thailand
(24) Kumasi Center for Collaborative Research in Tropical Medicine (KCCR), Ghana
(25) Kumasi Metropolitan Assembly (KMA), Ghana
(26) Kwame Nkrumah University of Science and Technology (KNUST), Ghana.
(27) Mahaweli Authority of Sri Lanka (MASL), Sri Lanka.
(28) Mekong Wetlands Biodiversity Conservation and Sustainable Use Programme (MWBP), Lao PDR
(29) Ministry of Agriculture and Water Resources of Uzbekistan (MAWR), Uzbekistan
(30) Ministry of Agriculture, Kazakhstan
(31) Ministry of Agriculture, Kyrgyzstan
(32) Ministry of Agriculture, Tajikistan
(33) Ministry of Agriculture, Turkmenistan
(34) Ministry of Food & Agriculture (MoFA), Ghana
(35) Ministry of Water Resources and Meteorology (MOWRAM), Cambodia
(36) National Agricultural and Forestry Research Institute (NAFRI), Lao PDR
(37) National Drainage Program (NDP), Pakistan
(38) National Institute for Soils and Fertilizers (NISF), Vietnam
(39) National Salinity Research Center (NSRC), Iran
(40) Nile Basin Initiative (NBI, NBI-ENTRO), Uganda/Ethiopia
(41) Pakistan Agricultural Research Council (PARC), Pakistan
(42) Professional Assistance for Development Action (PRADAN), India
(43) Seed and Plant Improvement Institute (SPII), Iran
(44) Sir Ratan Tata Trust (SRTT), India
(45) Scientific Information Center of Interstate Commission on Water Coordination (SIC-ICWC),Uzbekistan
(46) Sokoine University of Agriculture, Tanzania
(47) Tarbiat Modarres University, Iran
(48) Tashkent Irrigation and Melioration Institute, Uzbekistan
(49) The World Conservation Union (IUCN)
(50) Unilever Sri Lanka Limited (USL), Sri Lanka
(51) University of Agriculture Faisalabad (UAF), Pakistan
(52) University of Ghana (UG), Ghana.

(53) University of KwaZulu Natal, South Africa
(54) University of Melbourne, Australia
(55) University of Peradeniya, Sri Lanka
(56) University of Zimbabwe, Zimbabwe.
(57) Vietnam Institute for Water Resources Research, Vietnam
(58) Water & Energy Commission Secretariat (WECS), Nepal
(59) Water Research Commission (WRC), South Africa
(60) Water Research Institute of the CSIR (WRI), Ghana
(61) Water Resources Commission (WRC), Ghana
(62) Wuhan University, China

Knowledge Sharing Partners and Collaborators

(1) Bellanet International Secretariat, International Development Research Centre (IDRC), Canada
(2) Foundation for Ecological Security (FES), India
(3) Dhriiti-The Courage Within, India
(4) N. M. Sadguru Water & Development Foundation, India
(5) Society for Promoting Participative Ecosystem Management (SOPPECOM), India
(6) Development Support Centre (DSC), India
(7) Asian Development Research Institute (ADRI), India
(8) Indian Institute of Technology, Roorkee, India
(9) People Science Institute (PSI), India
(10) Gujarat Institute of Development Research (GIDR), India
(11) Trust Consulting, India
(12) Swaraj Foundation, India
(13) Tamil Nadu Agricultural University, India
(14) AKSHARA Network for DSS, India
(15) Scientific and Production Center for Agriculture
(16) Uzbek Scientific and Research Institute of Plant Production
(17) Tashkent State Agrarian University, Uzbekistan
(18) Ministry of Communication, Ghana
(19) Ghana Agricultural Information Network System (GAINS), Ghana
(20) Ghana Community Radio Network,
(21) Ghana Radio Ada, Ghana
(22) Busy Internet, Ghana
(23) TV3, Ghana
(24) Ghana Information Network for Knowledge Sharing (GINKS), Ghana
(25) Soil Water Management Research Group (SWMRG), Tanzania
(26) Farmer Support Group (FSG), South Africa
(27) The International Development Enterprises (IDE), USA (Global) and India
(28) Mekelle University (MKU), Ethiopia
(29) WaterNet, Zimbabwe

Subject Index

IWA Publishing's authorised EU representative for General Product Safety Regulations is Diane D'Arras, 15 rue Duret, 75116 Paris, France, e-mail: safety@iwap.co.uk.

Printed and bound by CPI Group (UK) Ltd, Croydon, CR0 4YY
16/06/2026
02140373-0001